Assessment and Management of Radioactive and Electronic Wastes

Edited by Hosam El-Din Saleh

Published in London, United Kingdom

IntechOpen

Supporting open minds since 2005

Assessment and Management of Radioactive and Electronic Wastes
http://dx.doi.org/10.5772/intechopen.77830
Edited by Hosam El-Din Saleh

Contributors
Brian Gareau, Cristina Lucier, Yin Lu, Jie Yu, Johnson Okorhi, Douglason Omotor, Helen Adereni, Deepak Yadav, Pradeep Kumar, Hosam El-Din M. Saleh

First published in London, United Kingdom, 2020 by IntechOpen
IntechOpen is the global imprint of INTECHOPEN LIMITED, registered in England and Wales, registration number: 11086078, 7th floor, 10 Lower Thames Street, London, EC3R 6AF, United Kingdom
Printed in Croatia

British Library Cataloguing-in-Publication Data
A catalogue record for this book is available from the British Library

Additional hard and PDF copies can be obtained from orders@intechopen.com

Assessment and Management of Radioactive and Electronic Wastes
Edited by Hosam El-Din Saleh
p. cm.
Print ISBN 978-1-78985-117-5
Online ISBN 978-1-78985-118-2
eBook (PDF) ISBN 978-1-78985-303-2

For EU product safety concerns:
IN TECH d.o.o., Prolaz Marije Krucifikse Kozulić 3, 51000 Rijeka, Croatia, info@intechopen.com or visit our website at intechopen.com.

Meet the editor

Hosam El-Din Saleh is a Professor of radioactive waste management in the Radioisotope Department, Nuclear Research Center, Atomic Energy Authority, Egypt. He has been awarded his MSc and PhD degrees in Physical Chemistry from Cairo University. He is interested in studying innovative, economic, and environmentally-friendly techniques for the management of hazardous and radioactive wastes. Saleh has authored many peer-reviewed scientific papers, chapters and has been a book editor of different books published by international publishers. He serves as reviewer or editor for several international journals. He was awarded the Scientific Encouragement Award from Atomic Energy Authority (2013), Encouragement Prize in Advanced Technical Sciences from Academy of Scientific Research and Technology (2014), and was listed in Marquis Who's Who in the World® for several editions.

Contents

Preface

This book provides guidance on assessment and management of radioactive and electronic waste. It refers to those wastes classified as unwanted materials generated from nuclear and industrial activities. These wastes impose the need to modify or change the character of raw or primary materials available to support or sustain the disposal of radioactive and electronic waste or new reuse in case of electronic waste. The development and application of approaches and technologies that provide economic and safe management is an essential issue in the treatment and disposal of hazardous wastes.

The authors have summarized their experience and present advances in relevant fields related to assessing the management of these materials. The book contains five chapters, organized in three sections that cover important research aspects of hazardous waste management technologies. The first section is an introductory chapter prepared by the editor to present a brief background on the generation, composting, types, and management of hazardous waste.

The second section presents the biological assessment and remediation of hazardous wastes. It comprises two chapters that deal with the toxicity testing bioassay using *Aremetia* spp. as a biological model. The use of this model is widespread due to its advantages. The first chapter is prepared by Yin Lu and Jie Yu. The second chapter, Phytoremediation of hazardous radioactive wastes, is prepared by Deepak Yadav and Pradeep Kumar.

The third section presents recycling and disposal of electronic waste, where Lucier and Gareau present the handling and regulation of e-waste as both a hazardous waste stream and as a source of secondary raw materials in the past decade. Okorhi Johnson prepared the last chapter in this book and it covers conducting the wastes from Industrialized Nations: A Socio-economic inquiry on E-waste Management for the Recycling Sector in Nigeria.

The editor wishes to thank all the participants in this book for their valuable contributions and to Ms. Nina Kalinic Babic for her assistance in finalizing the work. Acknowledgment for the IntechOpen staff members responsible for the completion of this book and other publications for free visible knowledge.

Hosam El-Din Mostafa Saleh
Atomic Energy Authority of Egypt,
Cairo, Egypt

Section 1

Introduction

Chapter 1

Introductory Chapter: Hazardous Wastes

Hosam M. Saleh and Samir B. Eskander

1. Introduction

Hazardous wastes can be defined as materials and equipment generated due to either natural or various anthropogenic activities and spiked with hazard ingredients, which there is no further use as well. Therefore, hazardous wastes are materials, direct disposal of which can pose threats to man and his environment. They can be explosive, flammable, oxidizing, poisonous/infectious, radioactive, corrosive and/or toxic [1].

According to the Resource Conservation and Recovery Act (RCRA) [40C.F.R. 261.31-33], a hazardous waste can be defined as a spiked material that poses a substantial threat to human health and/or his environment when segregated, sorted, handled, treated, stored, transported and disposed of under improper as well as uncontrolled conditions. Moreover, as spiked material, it has the capability to cause or can contribute to elevate mortality or a rise in epidemic and dangerous illness.

Hazardous waste generation and accumulation are the most acute brain teaser within the last two centuries, opposing world attention and priority for decision-making. Since the industrial revolution started, the hazardous wastes problem caused great and broaden damage to man's Ecosystems, therefore, it becomes an issue of serious not only for national but also for international concern [2].

Department of Environment and Energy, Australian Government, prescribed hazardous waste as which has any of the following characteristics: explosive; flammable liquids/solids; poisonous, toxic, ecotoxic; infectious substances, clinical wastes; waste oils/water, hydrocarbons/water mixtures, emulsions; wastes from the production, formulation and use of resins, latex, plasticizers, glues/adhesives; wastes resulting from surface treatment of metals and plastics; residues arising from industrial waste disposal operations; wastes which contain certain compounds such as copper, zinc, cadmium, mercury, lead and other heavy metals and asbestos; household waste; or residues arising from the incineration of household waste [3].

However, the US Environment Protection Agency (EPA) summarized that into four characteristics [4]:

- *Ignitability* or something flammable
- *Corrosivity* or something that can rust or decompose
- *Reactivity* or something explosive
- *Toxicity* or something poisonous (EPA, USA, etc.)

Hazardous waste-generating facilities can be differentiated into categories in accordance with the monthly amount of hazardous waste delivered. There are three categories, viz. large-quantity generators (LQGs), small-quantity generators (SQGs) and conditionally exempt small-quantity generators (CESQGs). To be nominated as a LQG, facility should throw more than 1000 kg of hazardous waste per month. Small-quantity generators generate between 100 and 1000 kg per month, while the third category, namely, CESQG facility, delivered less than 100 kg of hazardous waste each month [5, 6].

The nomination of the most famous categorization and the classification of hazardous waste are those based on the source that generates this waste and which can be distinguished from industrial waste, arisen from various industrial facilities; radioactive wastes generated due to the applications of radioisotopes in different fields of our life; medical, and pharmaceutical wastes, that are collected from health care facilities (HCFs), and so on ….

Healthcare waste (HCW) can be defined as the total wastes which are generated from a healthcare facility and would comprise non-hazardous or general waste and hazardous HCW. Besides, it includes the identical types of waste arisen from minor and scattered sources; the non-hazardous HCW is nominated as waste that does not pose any particular biological, chemical, radioactive or physical threats to man or to the environment. This group of waste con is managing following the municipal waste management hierarchy. The hazardous health care wastes (HHCWs) are considered the most crucial part of waste generated from the healthcare facilities due to their dangerous impacts on human and his ecosystems.

The main generators of healthcare waste are hospitals and other health facilities; limited medical centres; clinical centres, laboratories and research centres; mortuary and autopsy centres; animal research and testing laboratories; blood banks and collection services; laboratories for medical analysis; and nursing homes for the elderly [7].

Between 75 and 90% of the wastes generated by healthcare facilities that mainly resemble domestic wastes, therefore, are denoted as "non-hazardous" or "general healthcare wastes." They are collected mostly from the administrative, kitchen and housekeeping functions at healthcare facilities and may also include unspiked packaging waste and waste generated during maintenance of healthcare facilities. The remaining 10–25% of HCW are considered as "hazardous healthcare wastes" and can pose extensive environmental and health threats [8].

It is worth to state that pharmaceutical waste is not onefold category of waste but many and variable; moreover the chemicals that constitute pharmaceutical dosage forms are complex and variable. Healthcare wastes comprise sharps; non-sharps; disposable syringes and plastic equipment; blood, body tissue and parts, patient's excretions, chemicals and pharmaceuticals; chemotherapy ingredients; medical devices; and empty solution bags, bottles and containers, in addition to radioactive materials. The hazardous HCW can be classified into the following waste main groups:

2. Infectious waste

This group of wastes is assumed to contain pathogens (or their toxins) in a concentration that can be disease sources to a host. This group includes discarded materials or equipment, used for the diagnosis, and treatment of disease that has been in contact with body fluids, e.g. dressings, swabs, nappies, blood bags, etc., in addition to liquid waste comprising faeces, urine, blood, sputum or lung secretions.

3. Anatomical waste

Anatomical waste is a pathological category of hazardous HCW and includes body organs and tissues. Whether they can be infected or not, anatomical wastes are denoted in most cases as potential infectious wastes.

4. Radioactive waste

The most commonly used radioisotopes in healthcare facilities (HCFs) are technetium mTc-99 and gadolinium Ga-68 in therapeutic generators and cobalt Co-60, iodine I-131 and iridium Ir-192 for diagnosis and treatment. Low-level radioactive wastes are mainly the waste category generated in HCFs due to the applications of radioisotopes.

5. Hazardous pharmaceutical waste

Hazardous pharmaceutical wastes are a part of HCW generated not only in hospitals and medical centres but also in pharmacy. They comprise contaminated, spilt, unused and expired pharmaceutical products, as well as drugs and vaccines, and in addition discarded items used in the handling such as bottles, vials and connect tubing.

An important item of this category is all the drugs and equipment used for the mixing and administration of cytotoxic drugs. Cytotoxic drugs are used in chemotherapy treatment for cancer.

6. Sharps

Sharps are considered the most dangerous and highly infectious wastes generated at HCFs. They include needles, some surgical tools, syringes, disposable scalpels, blades, etc. Those items can result in cuts and punctured wounds; therefore, they should be collected, packed and handled in an extremely safe, controlled and proper method in the generation points to ensure the safety of the working staff.

7. Highly infectious waste

Body fluids of patients, with highly infectious diseases, microbial cultures and highly infectious stocks constitute what is named as the highly infectious wastes in the HCW scheme and are generated, mainly, from medical analysis and research laboratory activities.

8. Genotoxic/cytotoxic waste

This group of waste is accumulated from drugs generally used in oncology or radiotherapy units. It has high hazardous mutagenic and/or cytotoxic impacts. Excretions of cytotoxic drug- or chemically treated patients, i.e., faeces, vomit or urine, must be included as genotoxic waste. In specialized cancer treatment facilities, the controlled and proper treatment and safe disposal should be followed strictly to avoid contamination of the surrounding environment.

9. Hazardous chemical waste

Chemical waste covers the discarded chemicals that are collected after the disinfecting procedures or cleaning processes and generated in solid, liquid or gaseous form. They can be hazardous, i.e., toxic, corrosive, flammable, etc., and should be handled, treated and disposed of following the stated issues. Otherwise, the nonexplosive residues or small contents of outdated products can be treated as infectious waste.

Bulk chemotherapy waste is, also, managed as hazardous chemical waste and must be collected in hazardous waste containers. Firm management hierarchy should be applied for treating all bulk chemotherapy agents as hazardous waste when discarded.

10. Waste with a high content of heavy metals

Waste streams that have high concentration of heavy metals and their derivatives pose threats to healthcare facility as potentially highly toxic materials, e.g. cadmium or mercury from thermometers or manometers. They are categorized as a sub-group of chemical waste but should be managed separately.

The improper management of the hazardous healthcare wastes puts the healthcare workers, waste handlers and the community under the threats of infections, toxic impacts and injuries including damage of the environment. It also provides possibility for the segregated disposable medical equipment, to be resoled and reused, before their disinfection and sterilization, which can be a serious source for of epidemic disease for surrounding ecosystem [9].

It is conspicuous that hazardous materials being sold by a store, as pharmaceuticals, are not a hazardous waste until they are become expired. In general, the generator has to clearly decide that the material is a waste. In this trend, materials that can be sent to a reverse distributor must still be managed as a product, till the reverse distributor decides to dispose of the item. Loose pills, partial drug packages without all of the information on them, etc. are counted as waste and must be properly handled by the generator and/or the reverse distributor. In other words, for pharmaceutical products that meet the definition of a hazardous waste, their segregation, sorting, handling, transportation, treatment and final disposal must be carried out under the controlled rules. However, when these drugs are disposed of by the consumer, known as "ultimate user," then they are not counted a hazardous waste, since they are categorized as the exempt household waste. On the other hand, expired or un-needed pharmaceuticals at healthcare facilities, e.g., hospitals, pharmacies, medical centres, clinics, or other places dealing with drugs as business, are required to be managed as hazardous waste.

11. Hazardous waste management hierarchy

The management of hazardous wastes is a system carried out in sequences aiming at avoiding the escape of the harmful components from the waste to the man's surrounding environment. This hierarchy usually starts with segregation of the hazardous waste and is terminated by its final disposal. The source reduction can be considered as an issue in the HCW management topics even it takes place at every point of any production [10].

The methodology of a proper hazardous waste management hierarchy includes the upcoming processes in consequences: segregation and sorting, treatment, stabilization and solidification, storage and then final disposal (**Figure 1**). The full goals of this hierarchy, however it performed; when and wherever it carried out, are keeping human and his ecosystem safe, clean and tidy, moreover not burden the coming generation the hazardous problem due to our achievement.

However, pharmaceutical wastes are considered as a category of the healthcare waste, even though the healthcare professionals, always, do not pay the adequate attention to their proper management. There are a number of misconceptions regarding the proper methods for segregating, handling and treatment and disposing of this waste, markedly, at the low-income countries. It is worth to state that high-income countries (HICs) generate nearly up to 0.5 kg of hazardous waste/hospital bed/day; on the other hand, low-income countries delivered, only, about 0.2 kg. Even so, it is rare to find the healthcare wastes being separated into hazardous or non-hazardous wastes in LICs.

The main aim for treating/managing hazardous healthcare wastes is to convert it into to less or non-hazardous materials and stabilize their infectious, toxic and/or radioactive components by various techniques of solidification and encapsulation.

Many treatment methods have been used for healthcare wastes aiming at minimizing the threats of their hazard components and/or reducing the volume of the waste before disposal. Incineration of waste is the most widely applied technique for treatment of HCWs [11]. To avoid the disadvantages of incineration process, alternative methods have been applied such as pyrolysis [12], microwaving [13],

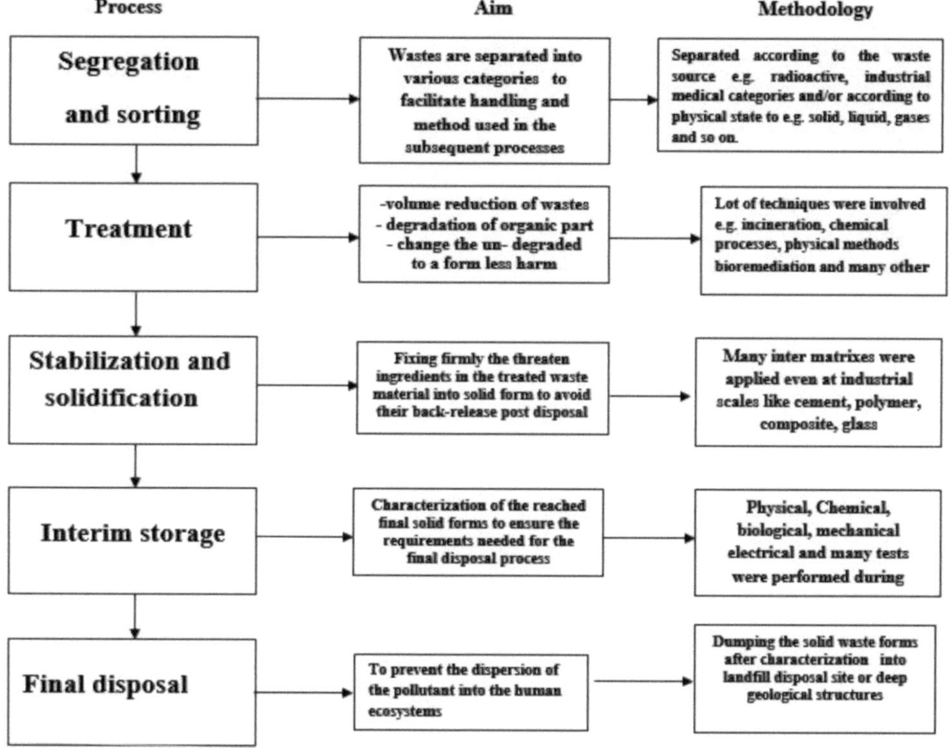

Figure 1.
Diagram for hazardous waste management hierarchy.

sterilization [14], steam treatment [15], thermal processing [16], wet oxidation [17] and many others.

The disposal of untreated or treated and solidified healthcare wastes must be undertaken in well-constructed landfills and in proper ways to eliminate the probability of the contamination of drinking, surface and groundwater.

12. Recommendation

Of the whole segregated waste generated by healthcare activities, nearly 85% is non-hazardous waste; the remaining 15% is regarded as hazardous material that can be infectious, toxic or radioactive. The terrible risks imposed by the unsafe, uncontrolled and improper management hierarchy, including the disposal of healthcare wastes in general, and their hazardous category, definitely, have long been approved all over the world. Insignificant and imperfect management of the healthcare waste puts healthcare human resources, waste handlers and transporters, and moreover the surrounding ecosystems, under the threats of infections, toxic impacts and damages.

Therefore, more researches, studies and efforts have to be undertaken through the World Health Organization (WHO) to raise awareness of the problem creating national and international action plans and to find the solutions, especially in the low-income countries. The WHO has to promote aids on the basis of the well-known five topics: management, training, regulatory and financial issues as well as technologies [18].

Author details

Hosam M. Saleh* and Samir B. Eskander
Radioisotope Department, Nuclear Research Center, Atomic Energy Authority, Egypt

*Address all correspondence to: hosamsaleh70@yahoo.com

References

[1] Muralikrishna IV, Manickam V. Hazardous waste management. In: Environmental Management: Science and Engineering for Industry. India Butterworth-Heinemann; 2017. pp. 463-494. DOI: 10.1016/B978-0-12-811989-1.00017-8

[2] Orloff K, Falk H. An international perspective on hazardous waste practices. International Journal of Hygiene and Environmental Health. 2003;**206**:291-302

[3] Hazardous Waste Act (Regulation of Exports and Imports). Department of the Environment and Energy, Australian Government; 1989

[4] U.S. EPA Test Methods, SW-846 Methods: 1010, 1020, and 1030, respectively. Available from: http://www.dtsc.ca.gov/LawsRegsPolicies/Title22/upload/OEARA_REG_Title22_Ch11_Art3.pdf

[5] Rosenfeld PE, Feng LGH. The biggest generators of hazardous waste in the US. In: Risks of Hazardous Wastes. UK: William Andrew; 2011

[6] Practice Greenhealth. Managing pharmaceutical waste. A 10-step blueprint for healthcare facilities in the United States. [Accessed: August 2008]

[7] WHO/UNICEF. Water, Sanitation and Hygiene in Health Care Facilities: Status in Low- and Middle-Income Countries. Geneva: World Health Organization; 2015

[8] World Health Organization (WHO). Safe Management of Wastes from Health-Care Activities. Geneva: Blue Book Second Edition; 2014

[9] Paudel R, Pradhan B. Health care waste management practice in a hospital. Journal of Nepal Health Research Council. 2010;**8**(2):86-90

[10] Vallero DA. Hazardous wastes. In: Waste: A Handbook for Management Academic Press, USA; 2011. pp. 393-423. DOI: 10.1016/B978-0-12-381475-3.10027-0

[11] Lerner BJ. Method for minimizing environmental release of toxic compounds in the incineration of wastes. 1997. United States Patent 5,607,654

[12] Eshleman RD. Sloped-bottom pyrolysis chamber and solid residue collection system in a material processing apparatus. 1995. United States Patent 5,417,170

[13] Kameda T, et al. Method of an apparatus for treating infectious medical wastes. 1994. United States Patent 5,322,603

[14] Kantor SL et al. Medical waste treatment unit. 2008. United States Patent 7,361,303

[15] Pappas CA. Medical waste disposal system. 1994. United States Patent 5,348,235

[16] Stevers PH et al. Waste materials processing apparatus and method. 2001. United States Patent 6,176,188

[17] Ghattas NK, Ghattas KM. Method and system for the decomposition of spent ion-exchange resins. 1992. German Patent DE 39 26 252 C2

[18] WHO. Status of Health-Care Waste Management in Selected Countries of the Western Pacific Region. Geneva: World Health Organization; 2015

Section 2

Biological Assessment and Remediation of Hazardous Wastes

Chapter 2

A Well-Established Method for the Rapid Assessment of Toxicity Using *Artemia* spp. Model

Yin Lu and Jie Yu

Abstract

Rapidly, relevantly, and efficiently toxicity assessment is the basis of continuous investigation and control of environmental contaminants. *Artemia* sp. is usually used as a biological model in cost-efficient bioassays under laboratory conditions to determine toxicity based on its advantageous properties of rapid hatching, easy accessibility, and sensitivity to toxic substances. The three sensitive endpoints of acute mortality, acute cyst hatchability, as well as behavioral response (such as swimming speed) are commonly used as evaluation criteria. The establishment of international standards for toxicity assessment of *Artemia* spp. is necessary. Further research is needed to obtain valuable insights from a biological perspective and for bio-conservation purposes.

Keywords: *Artemia*, toxicity assessment, mortality, hatchability, swimming speed

1. Introduction

Toxicology is the science of researching on the negative effects that chemical or physical agents may exert on living organisms under particular exposure conditions. It is a science that attempts to evaluate all the hazards, such as molecular toxicity, cytotoxicity, organ toxicity, etc., that are associated with a substance, as well as to quantitatively determine the exposure conditions under which these hazards or toxicities are induced [1, 2]. Additionally, toxicology is the science that studies the occurrence, character, frequency, mechanism, and risk elements associated with the adverse effects of toxic substances [2].

Many biological models can be applied for toxicity evaluation. Cell culture system is often used in vitro because it is economical and time-saving. But it is very difficult to infer the health of the whole organism, including humans, only from the results of in vitro cell tests. On the contrary, in vivo studies may provide improved prediction of biological reactions in intact systems (whole animal) but are generally expensive, time-consuming, and often elaborate, requiring extensive facilities and infrastructure [3]. Zebrafish (*Danio rerio*), as a classical model vertebrate organism, offers many practical advantages that can overcome these limitations to be highly suitable for application in toxicologically relevant research. Zebrafish can be employed as an outstanding in vivo model system to evaluate biological reactions and is a powerful platform to analyze in detail the mechanisms by which substances induce specific biological responses. Further, conditions in high-order vertebrates can be inferred from the results obtained using zebrafish because there is a remarkable similarity in cellular structure, signaling processes, anatomy, and physiology, particularly in the early stages of development [4–8]. Current estimates show that more than 90% of the human open reading frames are homologous to those in the genes of this fish [9]. Thus, investigations using this model system can reveal subtle interactions that are likely to be conserved across species.

2. Toxicity assessment with *Artemia* spp. and its advantages

The predominant EU Registration, Evaluation, Authorisation and Restriction of Chemicals (REACH) legislation with the aim of sound management of the eco-environment and protection of human societies promoted the decrease in the use of vertebrates and encouraged the use of invertebrates and plants, as well as organ, tissue, and cell cultures, as alternative study materials for toxicity and ecotoxicity testing [10]. Among various invertebrates screened and assessed to investigate their sensitivity to several physical and chemical substances, brine shrimps, *Artemia* spp., which are extremely sensitive to toxicity, stand out as one of the most frequently used species for toxicity testing [11] and are recognized and listed by the US Environmental Protection Agency [12] as the model organism for toxicity testing and emission monitoring.

Artemia sp. is a crustacean adapted to harsh conditions such as those in hypersaline lakes [13], living mainly on phytoplankton [14, 15]. It is closely related to other zooplankton such as copepods and daphnia (**Figure 1**) [16]. Normally, it is routinely employed as a test organism for ecotoxicological studies. The molecular, cellular, and physiological states of *Artemia* spp. change dramatically when they are under contamination stress [17]. At present, a variety of toxicity tests with *Artemia* spp. have been carried out covering both short-term acute and long-term chronic methods (**Table 1**), with the former being the more frequently used. Acute toxicity tests, which are highlighted in this paper, mainly assess the effect exposure to relatively high concentrations (at a mg/L level) for no more than 4 days (96 h). Toxicity under normal conditions is expressed as the lethal concentration causing the death of half of the tested animals (LC_{50}) and is also manifested in impeded hatching and swimming behavior. Chronic toxicity tests mainly have to do with the long-term exposure to relatively low concentrations (at a μg/L level) ranging from a few weeks up to the entire life cycle of *Artemia* spp. [18].

Figure 1.
An adult of Artemia spp.: male (left) and female (right).

Test type	Method	Parameter index
Short-term	Biomarker	AChE
		HSP
		Fluotox
		LP, TBARS, and TRed
		GRed, GPx, and GST
		ALDH and ATPases
	Hatching	Dry biomass
		Morphological disorder
		Size
		Teratogenicity
	Swimming	Speed
		Path length
	Immobilization	Mortality
Long-term	Growth	Body size
		Weight
		Morphological disorder
	Reproduction	Mating
		Reproductive rate
		Offspring
	Immobilization	Mortality

PS: AChE = acetylcholinesterase; HSP = heat stress proteins; LP = lipid peroxidation; TBARS = thiobarbituric acid reactive substances; TRed = thioredoxin reductase; GPx = glutathione peroxidase; GST = glutathione S-transferase; GRed = glutathione reductase; ALDH = aldehyde dehydrogenase; and ATPases = adenyltriphosphatase

Table 1.
Summary of Artemia short- and long-term toxicity tests [19].

Considering the environmental aspect, *Artemia* spp. nauplii were employed to assess the toxicity of various hazardous metal substances such as As, Cr, Sn, etc. [19–22]; organic compounds including pharmaceuticals, agrichemicals, etc. [23–26]; and environmental media such as wastewater [27], seawater [28], and marine discharges [29].

The principal advantages of using *Artemia* spp. in toxicity testing are as follows: (1) rapidity in hatching, (2) cost-efficiency, and (3) commercial availability of nauplii hatched from durable cysts, which dispenses with the need for self-culturing [30, 31]. Moreover, other significant factors that have been taken into consideration include good cognition of its biological and ecological features, small size allowing for easy laboratory operation, as well as its well-developed adaptability to diversified testing conditions [30, 32]. It is noteworthy that the complex adaptive response evolved by *Artemia* to live through and thrive in critical conditions not only explains why it is a favorable candidate for toxicity testing but to some extent also offers insights with regard to biological and environmental perspectives, which in turn might contribute to toxicity testing itself and eventually the well-being of human populations. With that being said, the response mechanism developed by *Artemia* to deal with harsh conditions [13] is worth mentioning (see **Figures 2** and **3**). The harsh living condition is exemplified in hypersaline lakes (salty lakes) where *Artemia* is often the only macroplanktonic inhabitant [13]. The survival and reproduction of

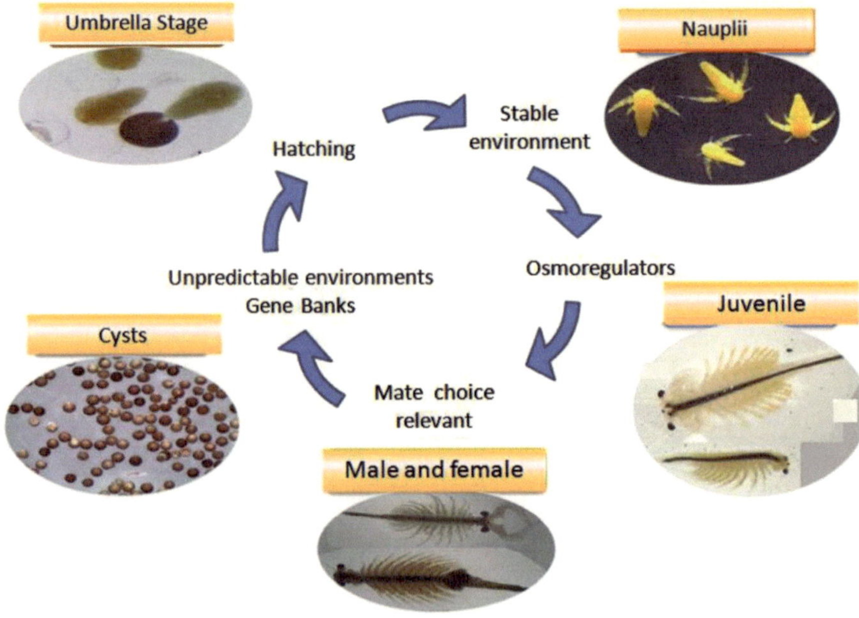

Figure 2.
The life cycle and different stages of Artemia as a salty survivor.

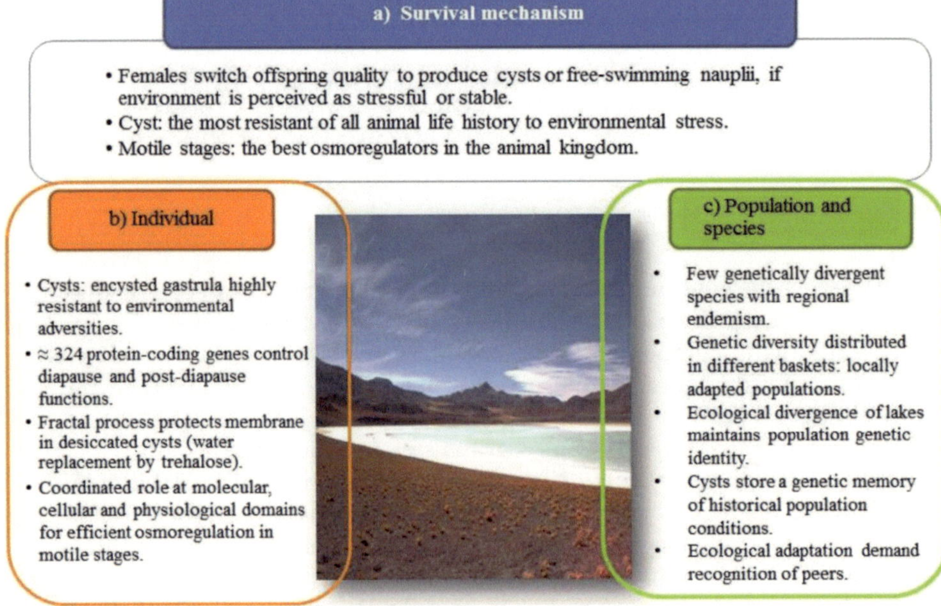

Figure 3.
The reproduction of Artemia brine shrimp (individuals, populations, and species) subject to critical life conditions imposed by salty lakes.

the brine shrimp *Artemia* (individuals, populations, and species) subject to critical life conditions imposed by salty lakes, as schemed in **Figures 2** and **3**, may be summarized as follows: (1) Females are able to cope with the forthcoming environmental conditions by switching the type of offspring to produce either cysts under stressful conditions or free-swimming nauplii under stable conditions, and (2) cysts are the most environmental stress-resistant among all animal life history forms, while motile stages are the best osmoregulators in the animal kingdom [33]. Cysts are gene banks that store a genetic memory of historical population conditions. They play a role aiding in the dispersal of *Artemia* and serve as reservoirs of genetic variability [34] and the source of evolutionary change and resilience.

3. Application status of the toxicity assessment with *Artemia* spp.

Ecotoxicological studies employing *Artemia* spp. as testing species have been extensively performed, and among the endpoints that were mainly investigated, acute mortality, acute cyst hatchability, as well as behavioral response, as a result of their relatively high sensitivity, are commonly used.

3.1 Acute mortality test

Acute mortality is one of the most commonly used endpoints for toxicity testing, though there is no standardized protocol based on OECD and ISO regulations. Since the establishment of the *Artemia* Reference Center (ARC-test) and the issuance of the first short-term acute mortality (24 h static test) protocol with *Artemia* larvae [35–38], extensive toxicity assessment research using this bioassay has been carried out via calculating the median effectiveness concentration on mortality (24 h LC_{50}). Besides observation of lethal endpoints for *Artemia* exposed to reference toxicants including $CuSO_4$, $K_2Cr_2O_7$, and SDS [39, 40], many are related with toxicity monitoring of environmental pollutants such as heavy metals, pesticides, oil drilling fluids, organic compounds of ecotoxicological concern, and others [41–44]. Indeed, in the wake of various environmental issues challenging humans and living surroundings, the importance of toxicity assessment using *Artemia* has been gradually recognized and more frequently employed. The following are two examples in recent years.

The “Brine Shrimp Lethality” study is one of the biological assays to determine the safe exposure limit of naturally occurring agents extracted from plants before being used as pesticides for crops and for other botanical protections [16]. Crop protection is one of the important food safety-related issues and is thus vital to human populations worldwide. As crop protection nowadays rely heavily on synthetic pesticides [45], the massive use of these pesticides for the purpose of killing pests and preventing diseases in plants has inevitably led to several side effects such as pest resistance resulting in the use of increased application rates [46], harm to nontarget organisms, and environmental contaminations with the potential influence on the food chain [47] that might cause pesticide poisoning of humans directly. Botanically derived natural products therefore have attracted attention among phytochemists. “Brine Shrimp Lethality,” a rapid general bioassay, offers a unique advantage in the standardization and quality control of those bioactive compounds that are usually undetectable using traditional physical analytical methods. The objective of carrying out the biological assay focuses on establishing a cause-effect relationship (**Figure 4**) between exposure to a hazardous substance and an appeared effect expressed by dose-response curve to determine a safe exposure limit [48]. The threshold level as well as the toxicity features obtained from the dose-response curves can help determine the safe levels of chemicals in botanical extracts and chemical exposure [49]. The threshold information ($ThD_{0.0}$) measured in mg/kg/day and based on the assumption that human beings are as sensitive as the tested animals; in this case the brine shrimp *Artemia* sp. is of paramount importance in generalizing animal data to humans and interpolating what might be considered a safe human dose for a given chemical.

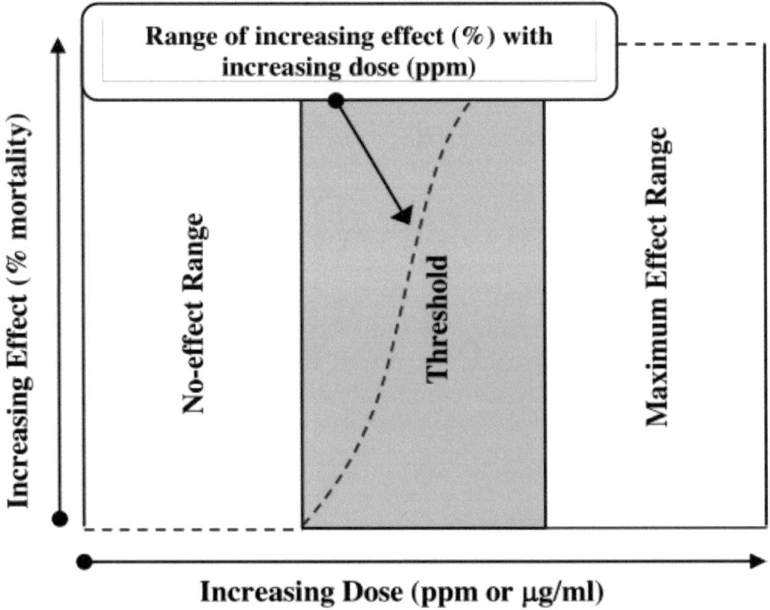

Figure 4.
Dose-response curve.

Another example in relation to the *Artemia* acute toxicity test [50, 51] is for the purpose of prevention and reduction of red tides. The red tide induced by algae is quite disastrous and may pose a threat to inshore fishery. The poisonous *Chattonella marina* that produces reactive oxygen species (ROS) [52] and hemolytic toxins [53] is one kind of red tide-related algae and has caused massive fish death and a considerable amount of economic loss in many places around the world. The "Brine Shrimp Lethality" study in this regard can help reveal the toxic characteristics of *Chattonella marina*, offer some valuable red tide prevention evidences, and further benefit the offshore fishery industry.

3.2 Acute cyst hatching test

Analogous to the acute mortality test, acute cyst hatching testing, which observes the retarded emergence of nauplii from cysts [54] or the morphological disorders and size of hatched nauplii [55] when exposed to toxic agents, is another frequently used assay for toxicity assessment. The hatching toxicity test lasting between 24 and 96 h in static conditions was investigated to assess the effect of environmentally deleterious agents such as heavy metals [54, 56, 57], organic compounds [58, 59], antibiotic drugs [60], and others. As temperature profoundly influences the hatching percentage of cysts [61] and significantly affects the chemicals' effect [62], it is a variable of great interest to be considered while carrying out the hatching test, and the use of a full temperature range might help increase the ecotoxicological data in an extensive manner.

3.3 Acute behavioral test (swimming speed)

Regarding the acute behavioral test, motion behavior changes in response to pollutant exposure have been investigated for a range of aquatic organisms [63–67]. In particular, swimming speed as a sublethal behavioral endpoint can be detected by employing a video camera tracking system developed by Faimali et al. [63], also known as the Swimming Speed Alteration (SSA) recording system, which has already been used on the brine shrimp, *Artemia* [68]. Moreover, the research results of Garaventa et al. [68] and Manfra et al. [69] showed that swimming speed was more sensitive than mortality and had a sensitivity similar to and sometimes higher than that of the hatching rate endpoint. Therefore, it is a well-defined behavioral response and an adaptable endpoint that can be used for ecotoxicity testing. For instance, Manfra et al. [69] recorded the swimming speed alteration of *Artemia* exposed to diethylene glycol (DEG), an organic substance ecotoxicological concern, and observed a decline in the swimming speed under the toxicant concentration of 40–160 g/L after 24 h exposure and 10–160 g/L after 48 h exposure. Another example is related with marine pollution such as oil spilling, oil mining, and oily water discharge that can greatly threaten human health as contaminants can be accumulated in the human body through the food chain. In this regard, *Artemia* spp., as one of the toxicity-monitoring species, is of great importance in the evaluation of the health of the marine ecosystem. Pan [70] investigated the swimming speed and motion angle alteration of *Artemia* exposed to diesel oil. For comparison purposes, when experiments were carried out under normal conditions, namely, seawater, the swimming speed of *Artemia* increased by 51%, from 2.47 mm/s at the start time to 3.72 mm/s after 12 h exposure on average, and in a similar trend, the motion angle of *Artemia* increased from 25 to 37°. In contrast, when subject to diesel oil, the swimming speed of *Artemia* decreased by 40%, from 2.37 mm/s at the start time to 1.42 mm/s after 12 h exposure on average, and in a similar trend, the motion angle of *Artemia* decreased from 30 to 21°.

4. Prospects for development of toxicity assessment with *Artemia* spp.

To rapidly figure out the deleterious effects brought about by environmental toxicants, acute toxicity assessment with *Artemia* spp. is of paramount importance as it shows a decent ability in pre-screening of toxic substances [10] and, thus, will be further developed in the future.

Despite the widespread application of this bioassay, there is currently no internationally standardized method. Hence, intercalibration exercises as well as international standardization activities are rather necessary [71]. Among the three frequently used endpoints involving acute mortality, acute cyst hatchability, as well as behavioral response, acute mortality was intercalibrated based on the available standards [40, 69, 72], while acute hatchability was intercalibrated at the Italian level [69]. To make *Artemia* spp. an international standard model in ecotoxicity testing calls for joint efforts engaging all relevant stakeholders including the government, NGOs, researchers, industry, consumer associations, and others.

Swimming speed as the most popular behavioral endpoint promises to be of great potential. This is because results can be obtained via easy video camera analysis at ease and also because the swimming speed is of great ecological significance as the behavior alteration means an integral whole body response that can connect the physiological and ecological features of an organism with its environment [73]. Nevertheless, to better employ this endpoint, the interaction of *Artemia* spp. with contaminants, particularly the mechanisms of response to toxic effect, needs to be illuminated.

One is to believe that owing to the advantages of using *Artemia* spp. as the biological model described in the previous section of this paper, besides toxic testing application itself, application into other environmentally related fields such as applied biology might also be put into practice. For example, from a bio-conservation point of view, the unique biological characteristics of brine shrimp *Artemia* make it a model organism to evaluate management policies for the protection of aquatic resources [74]. *Artemia* is such a versatile creature that it is a paradigmatic model having not only scientific research values but also the ability to satisfy human needs, owing to its unique life traits including a well-developed adaptability to high salinity conditions as well as easy handling under laboratory conditions, which have been successfully applied to marine fish farming that uses *Artemia* nauplii as food for fish larvae. However, the booming marine fish farming activity worldwide is likely to give rise to some risks in terms of the high genetic divergence between different *Artemia* species. Exploitation of new *Artemia* cyst harvesting sites and introduction of an exotic species linked to traits relevant to aquaculture can drive other local genotypes to extinction. Risk assessment and evaluation of management decisions in exploited resources, for instance, the availability of genetic information as well as molecular tools for follow-up gene pool monitoring, therefore, become quite necessary in order to maintain biodiversity. Gene banks established from cysts collected from various sites guarantee population persistence while proceeding with management affairs. Taking into account the simple constitution of hypersaline habitats, the evaluation of population/species persistence with *Artemia* can be modeled in laboratories and further extrapolated to other species, offering some of the aspects regarding rational aquatic resource utilization and, more importantly, biodiversity preservation.

5. Conclusions

After more than five decades of use in ecotoxicology, *Artemia* spp. have demonstrated its suitability for use in pre-screening of toxic agents [10]; thus, it seems that *Artemia* sp. endpoints may be used as a toxicity testing method to meet market demand, even though there are no internationally standardized toxicity testing protocols at present according to the ISO and OECD.

Biomarkers and teratogenicity are the less popular endpoints used in short-term toxicity tests because of their limited sensitivity. However, behavioral endpoints, especially swimming inhibition, seem to have a wider application potential in the future with the development of computer technology. Both continuous and intermittent observations of single or groups of living organisms can be studied by image and video analysis. Hatching rate and acute mortality are the most commonly used endpoints in the standardization process at a different level. Usually, hatching rate (48 h static test) was intercalibrated at the Italian level [69], while acute mortality (24 h static test) was intercalibrated based on the available standard [40] at the Italian [69] as well as the European level [72]. Both provided data on $CuSO_4$ as a reference toxicant. Among the long-term toxicity tests, the 14-d static renewal mortality test was intercalibrated at the Italian level [69] with SDS according to the UNICHIM protocol (2012).

Further concentrated efforts are necessary to make *Artemia* sp. an official internationally recognized standard biological model in ecotoxicology evaluation. It involves (I) a national member (who then contacts the ISO) upon a request by an industry sector or group for a standard; (II) scope, main definitions, and contents of standards which are scientifically assessed by experts in relevant fields; and (III) multi-stakeholder discussion and reviewing process including experts from related industries, consumer associations, academic institutions, nongovernmental organizations, and governments.

Acknowledgements

The authors of this study express their gratitude to the National Natural Science Foundation of China (No. 31600257), Public Projects of Zhejiang Province

(No. 2016C32022), Academic Climbing Project for Young and Middle-Aged Leads in Universities of Zhejiang Province (pd2013339), and Project of Zhejiang Provincial Department of Education (Y201738582) for their financial supports of this study.

Author details

Yin Lu* and Jie Yu
College of Biology and Environmental Engineering, Zhejiang Shuren University, Hangzhou, China

*Address all correspondence to: luyin_zjsru@aliyun.com

References

[1] Gosselin RE, Smith RP, Hodge HC. Clinical Toxicology of Commercial Products. 5th ed. Baltimore, MD: Williams & Wilkins Company; 1984. 217 p. DOI: 10.1016/S0022-3476(57)80185-1

[2] Williams PL, James RC, Roberts SM. Principles of Toxicology: Environmental and Industrial Applications. 2nd ed. London: John Wiley & Sons; 2003. 325 p. DOI: 10.1016/S0160-4120(00)00083-0

[3] Akimenko MA, Johnson SL, Westerfield M, Ekker M. Differential induction of four msx homeobox genes during fin development and regeneration in zebrafish. Development. 1995;**121**(2):347-357. DOI: US8114972 B2

[4] Aparicio S, Chapman J, Stupka E, Putnam N, Chia JM, Dehal P, et al. Whole-genome shotgun assembly and analysis of the genome of *Fugu rubripes*. Science. 2002;**297**(5585):1301-1310. DOI: 10.1126/science.1072104

[5] Blechinger SR, Warren JT, Kuwada JY, Krone PH. Developmental toxicology of cadmium in living embryos of a stable transgenic zebrafish line. Environmental Health Perspectives. 2002;**110**(10):1041-1046. DOI: 10.1289/ehp.021101041

[6] Busquet F, Nagel R, von Landenberg F, Mueller SO, Huebler N, Broschard TH. Development of a new screening assay to identify proteratogenic substances using zebrafish danio rerio embryo combined with an exogenous mammalian metabolic activation system (mDarT). Toxicological Sciences. 2008;**104**(1):177-188. DOI: 10.1093/toxsci/kfn065

[7] Harper SL, Dahl JL, Maddux BLS. Proactively designing nanomaterials to enhance performance and minimize hazard. International Journal of Nanotechnology. 2008;**5**(1):124-142. DOI: 10.1504/ijnt.2008.016552

[8] Henken DB, Rasooly RS, Freeman N, et al. Recent papers on zebrafish and other aquarium fish models. Zebrafish. 2003;**1**:305-311. DOI: 10.1089/zeb.2005.2.55

[9] Kimmel CB, Ballard WW, Kimmel SR, Ullmann B, Schilling TF. Stages of embryonic development of the zebrafish. Developmental Dynamics. 1995;**203**(3):253-310. DOI: 10.1002/aja.1002030302

[10] Dvorak P, Benova K, Vitek J. Alternative Biotest on *Artemia franciscana*. In: Begum G, editor. Ecotoxicology. Rijeka, Croatia: InTech; 2012. pp. 51-74. DOI: 10.5772/29114

[11] Van Steertegem M, Persoone G. Cyst-based toxicity tests: V. Development and critical evaluation of standardized toxicity tests with the brine shrimp (Anostraca, Crustacea). In: Soares AMVM, Calow P, editors. Progress in Standardization of Aquatic Toxicity Tests. New York: Lewis Publishers; 1993. pp. 81-97. DOI: 10.1016/j.toxicon.2012.12.024

[12] United states environmental protection agency (US EPA). Proceedings of seminar on methodology or monitoring the marine environment. USE PA EPA600/4-74-004; 1983

[13] Gajardo GM, Beardmore JA. The brine shrimp *Artemia*: Adapted to critical life conditions. Frontiers in Physiology. 2012;**3**:1-8. DOI: 10.3389/fphys.2012.00185

[14] Persoone G, Sorgeloos P. General aspects of the ecology and biogeography of Artemia. In: Persoone G, Sorgeloos P, Roels PO, Jaspers E, editors. The Brine Shrimps *Artemia*, 3. Ecology, Culturing, Use in Aquaculture. Belgium: Universa Press; 1980. pp. 3-24

[15] Triantaphyllidis GV, Abatzopoulos TJ, Sorgeloos P. Review of the

biogeography of the genus *Artemia* (Crustacea, Anostraca). Journal of Biogeography. 1998;**25**:213-226. DOI: 10.1046/j.1365-2699.1998.252190.x

[16] Zubairi SI, Othman ZS, Sarmidi MR, Aziz RA. Environmental friendly bio-pesticide rotenone extracted from *Derris* sp.: A review on the extraction method, toxicity and field effectiveness. Jurnal Teknologi Sciences & Engineering. 2016;**78**:47-69. DOI: 10.11113/jt.v78.5942

[17] Marigómez I, Soto M, Orbea A, Cancio I, Cajaraville MP. Biomonitoring of environmental pollution along the Basque coast, using molecular, cellular and tissue-level biomarkers: An integrative approach. In: Borja A, Collins M, editors. Oceanography and Marine Environment of the Basque Country. Amsterdam: Elsevier; 2004. pp. 335-364. DOI: 10.1016/S0422-9894(04)80052-7

[18] Pane L, Agrone C, Giacco E, Somà A, Mariottini GL. Utilization of marine crustaceans as study models: A new approach in marine ecotoxicology for European (REACH) regulation 91. In: Ghousia B, editor. Ecotoxicology. Rijeka, Croatia: InTech; 2012. p. 146. DOI: 10.5772/29114

[19] Libralato G, Prato E, Migliore L, Cicero AM, Manfra L. A review of toxicity testing protocols and endpoints with *Artemia* spp. Ecological Indicators. 2016;**69**:35-49. DOI: 10.1016/j.ecolind.2016.04.017

[20] Brix KV, Cardwell RD, Adams WJ. Chronic toxicity of arsenic to the great salt Lake brine shrimp, *Artemia franciscana*. Ecotoxicology and Environmental Safety. 2003;**54**:169-175. DOI: 10.1016/S0147-6513(02)00054-4

[21] Leis M, Manfra L, Taddia L, Chicca M, Trentini P, Savorelli F. A comparative toxicity study between an autochthonous *Artemia* and a non-native invasive species. Ecotoxicology. 2014;**23**(6):1143-1145. DOI: 10.1007/s10646-014-1252-4

[22] Hadjispyrou S, Kungolos A, Anagnostopoulos A. Toxicity, bioaccumulation, and interactive effects of organotin: Cadmium and chromium on *Artemia franciscana*. Ecotoxicology and Environmental Safety. 2001;**49**:179-186. DOI: 10.1006/eesa.2001.2059

[23] Xu X, Lu Y, Zhang D, Wang Y, Zhou X, Xu H, et al. Toxic assessment of triclosan and triclocarban on *Artemia salina*. Bulletin of Environmental Contamination and Toxicology. 2015;**95**:728-733. DOI: 10.1007/s00128-015-1641-2

[24] Kuwabara K, Nakamura A, Kashimoto T. Effect of petroleum oil, pesticides, PCBs and other environmental contaminants on the hatchability of *Artemia salina* dry eggs. Bulletin of Environmental Contamination and Toxicology. 1980;**25**:69-74. DOI: 10.1007/BF01985489

[25] Varó I, Taylor AC, Ferrando MD, Amat F. Effect of endosulfan pesticide on the oxygen consumption rates of nauplii of different Spanish strains of *Artemia*. Journal of Environmental Science and Health—Part B Pesticides, Food Contaminants, and Agricultural Wastes. 1997;**32**:363-375. DOI: 10.1080/03601239709373092

[26] Varó I, Navarro JC, Amat F, Guilhermino L. Characterisation of cholinesterases and evaluation of the inhibitory potential of chlorpyrifos and dichlorvos to *Artemia salina* and *Artemia parthenogenetica*. Chemosphere. 2002;**48**:563-569. DOI: 10.1016/S0045-6535(02)00075-9

[27] Krishnakurmar PK, Dineshbabu AP, Sasikumar G, Bhat GS. Toxicity evaluation of treated refinery effluent using brine shrimp (*Artemia salina*) eggs and larval bioassay. Fishery Technology. 2007;**44**:85-92. URI: http://eprints.cmfri.org.in/id/eprint/5791

[28] Manfra L, De Nicola E, Maggi C, Zambianchi E, Caramiello D, Toscano A, et al. Exposure of rotifers, crustaceans and sea urchins to produced formation waters and seawaters in the Mediterranean Sea. Journal of the Marine Biological Association of the UK. 2011;**91**(1):155-161. DOI: 10.1017/s0025315410001037

[29] Manfra L, Maggi C, Bianchi J, Mannozzi M, Faraponova O, Mariani L, et al. Toxicity evaluation of produced formation waters after filtration treatment. Natural Science. 2010;**2**(1):33. DOI: 10.4236/ns.2010.21005

[30] Nunes BS, Carvalho FD, Guilhermino LM, Van Stappen G. Use of the genus *Artemia* in ecotoxicity testing. Environmental Pollution. 2006;**144**: 453-462. DOI: 10.1016/j.envpol.2005.12.037

[31] Manfra L, Savorelli F, Pisapia M, Magaletti E, Cicero AM. Long-term lethal toxicity test with the crustacean *Artemia franciscana*. JoVE. 2012;**62**:2182-2185. DOI: 10.3791/3790

[32] Kokkali V, Katramados I, Newman JD. Monitoring the effect of metal ions on the mobility of *Artemia salina* nauplii. Biosensors. 2011;**1**(2):36-45. DOI: 10.3390/bios1020036

[33] Clegg JS, Trotman C. Physiological and Biochemical Aspects of *Artemia* Ecology. Netherlands: Kluwer Academic Publishers; 2002. 129 p. DOI: 10.1007/978-94-017-0791-6_3

[34] Gajardo G, Beardmore JA. Ability to switch reproductive mode in Artemia is related to maternal heterozygosity. Marine Ecology Progress Series. 1989;**56**:191-195. DOI: 10.3354/meps055191

[35] Gonzalo MG, John AB. The brine shrimp *Artemia*: Adapted to critical life conditions. Frontiers in Physiology: Front Physiol. 2012. 185 p. DOI: 10.3389/fphys.2012.00185

[36] Vanhaecke P, Persoone G, Claus C, Sorgeloos P. Proposal for a short-term toxicity test with *Artemia* nauplii. Ecotoxicology and Environmental Safety. 1981;**5**(3):382-387. DOI: 10.1016/0147-6513(81)90012-9

[37] Vanhaecke P, Persoone G. Report on an intercalibration exercise on a short-term standard toxicity test with *Artemia* nauplii (ARC-test). In: Leclerc H, Dive D, editors. Les tests de toxicité aigue en milieu aquatique, Les colloques de l'INSERM. Paris: Ministère de la Santé: Institut National de la Santé et de la Recherche Médicale; 1981. pp. 359-376

[38] Vanhaecke P, Persoone G. The ARC-test: A standardized short-term routine toxicity test with *Artemia* nauplii. Methodology and evaluation. In: Persoone G, Jaspers E, Claus C, editors. Ecotoxicological Testing for the Marine Environment. Experience Papers: Tests with Specific Groups of Organisms; Tests with Specific Chemicals; Tests Using a Specific Technology; Tests Studying Specific Effects; Case Study. Ghent: State University of Ghent and Institute of Marine and Scientific Research; 1984. pp. 143-158

[39] Guzzella L. Saggio di tossicità acuta con *Artemia* spp. Biologia Ambientale. 1997;**1**:4-9. (in Italian)

[40] APAT and IRSA-CNR. Metodi analitici per le acque (in Italian). 2003

[41] Zhang QT, Hu GK. Studies on joint toxicity of five heavy metal ions to *Artemia*. Journal of Tianjin University of Science and Technology. 2010;**25**(2):26-29, 44

[42] Wu ZF, Liu XG, Wang GX. Evaluating and modeling the toxicity of binary mixtures of heavy metals and organophosphate pesticides to *Artemia salina*. Asian Journal of Ecotoxicology. 2013;**8**(4):602-608

[43] Liu LP, Chu CY, Zhang QQ, Xue YZ, Li GR, Zhang Y. The application of

Artemia in the toxicity test of drilling fluid. Journal of Ocean University of China (English Edition). 2010;**40**(9):96-100

[44] Manfra L, Canepa S, Piazza V, Faimali M. Lethal and sublethal endpoints observed for *Artemia* exposed to two reference toxicants and an ecotoxicological concern organic compound. Ecotoxicology and Environmental Safety. 2016;**123**:60-64. DOI: 10.1016/j.ecoenv.2015.08.017

[45] Coats JR. Risks from natural versus synthetic insecticides. Annual Review of Entomology. 1994;**39**(1):489-515. DOI: 10.1146/annurev.en.39.010194.002421

[46] Stoll G. Natural Crop Protection in the Tropics–based on Local Farm Resources in the Tropics and Subtropics. Weikersheim: Josef Margraf; 1992. DOI: 10.12691/wjar-3-2-5

[47] Copping LG. The Biopesticide Manual: World Compendium. British Crop Protection Council; 1998

[48] Ottoboni MA. The Dose Makes the Poison; A Plain-Language Guide to Toxicology. Van Nostrand Reinhold; 1991

[49] Levine RR. Pharmacology: Drug Actions and Reactions. Boston: Little, Brown and Co.; 1973. 279 p. DOI: 10.1177/106002808401800741

[50] Saiful IZ, Zetty SO, Mohamad RS, Ramlan AA. Environmental friendly bio-pesticide rotenone extracted from *Derris* sp.: A review on the extraction method, toxicity and field effectiveness. Jurnal teknologl (Sciences & Engineering). 2016;**78**(8):47-69. DOI: 10.11113/jt.v78.5942

[51] Xu YH, Jiang T, Wang R, Shen PP, Wu N, Jiang TJ. Acute toxicity of *Chattonella Marina* on *Artemia Sinica*. Journal of Jinan University (Natural Sciencei). 2012;**33**(5):510-515. DOI: 10.3969/j.issn.1000-9965.2012.05.016

[52] Oda T, Nakamura A, Midori S. Generation of reactive oxygen species by Raphidophycean phytoplankton. Bioscience, Biotechnology, and Biochemistry. 1997;**61**:1658-1662. DOI: 10.1271/bbb.61.1658

[53] Kuroda A, Nakashima T, Yamaguichi K. Isolation and characterization of light-dependent hemolytic cytotoxin from harmful red tide phytoplankton *Chattonella marina*. Comparative Biochemistry and Physiologyl. PTC. 2005;**141**(3):297-305. DOI: 10.1016/j.cca.2005.07.009

[54] Bagshaw JC, Rafiee P, Matthews CO, MacRae TH. Cadmium and zinc reversibly arrest development of *Artemia* larvae. Bulletin of Environmental Contamination and Toxicology. 1986;**37**:289-296. DOI: 10.1007/BF01607763

[55] Neumeyer CH, Gerlach JL, Ruggiero KM, Covi JA. A novel model of early development in the brine shrimp, *Artemia franciscana*, and its use in assessing the effects of environmental variables on development, emergence, and hatching. Journal of Morphology. 2014;**276**:342-360. DOI: 10.1002/jmor.20344

[56] Brix KV, Gerdes RM, Adams WJ, Grosell M. Effect of copper, cadmium, and zinc on the hatching success of brine shrimp (*Artemia franciscana*). Archives of Environmental Contamination and Toxicology. 2016;**51**:580-583. DOI: 10.1007/s00244-005-0244-z

[57] Sarabia R, Del Ramo J, Diaz-Mavans J, Torreblanca A. Development and reproductive effects of low cadmium concentration on *Artemia parthenogenetica*. Journal of Environmental Science and Health. Part A, Toxic/Hazardous Substances & Environmental Engineering. 2003;**38**:1065-1071. DOI: 10.1081/ESE-120019864

[58] Alyürüc H, Çavas L. Toxicity of diuron and irgarol on the hatchability

and early stage development of *Artemia salina*. Turkish Journal of Biology. 2013;**37**:151-157. DOI: 10.3906/biy-1205-39

[59] Rotini A, Manfra L, Canepa S, Tornambè A, Migliore L. Can *Artemia* hatching assay be a (sensitive) alternative tool to acute toxicity test? Bulletin of Environmental Contamination and Toxicology. 2015;**95**:745-751. DOI: 10.1007/s00128-015-1626-1

[60] Migliore L, Civitareale C, Brambilla G, Dojmi di Delupis G. Toxicity of several important antibiotics to *Artemia*. Water Research. 1997;**31**:1801-1806. DOI: 10.1007/s00128-015-1626-1

[61] Vanhaecke P, Sorgeloos P. International study on *Artemia*. XLVII. The effect of temperature on cyst hatching: Larval survival and biomass production for different geographical strains of brine shrimp *Artemia* spp. Annales de la Société Royale Zoologique de Belgique. 1989;**119**:7-23. DOI: 10.1111/j.1440-1681.2008.05051.x

[62] Koutsaftis A, Aoyama I. Toxicity of diuron and copper pyrithione on the brine shrimp *Artemia franciscana*: The effects of temperature and salinity. Journal of Environmental Science and Health, Part A: Toxic/Hazardous Substances & Environmental Engineering. 2008;**43**:1581-1585. DOI: 10.1080/10934520802329794

[63] Faimali M, Garaventa F, Piazza V, Corrà C, Magillo F, Pittore M, et al. Swimming speed alteration of larvae of *Balanus amphitrite* as behavioural end-point for laboratory toxicological bioassays. Marine Biology. 2006;**149**(1):87-96. DOI: 10.1007/s00227-005-0209-9

[64] Rao JV, Kavitha P, Jakka NM, Sridhar V, Usman P. Toxicity of organophosphates on morphology and locomotor behavior in brine shrimp, *Artemia salina*. Archives of Environmental Contamination and Toxicology. 2007;**53**:227-232. DOI: 10.1007/s00244-006-0226-9

[65] Kienle C, Gerhardt A. Behavior of *Corophium volutator* (Crustacea, Amphipoda) exposed to the water-accommodated fraction of oil in water and sediment. Environmental Toxicology and Chemistry. 2008;**27**: 599-604. DOI: 10.1897/07-182

[66] Xuereb B, Lefevre E, Garric J, Geffard O. Acetylcholinesterase activity in *Gammarus fossarum* (Crustacea Amphipoda): Linking AChE inhibition and behavioural alteration. Aquatic Toxicology. 2009;**94**:114-122. DOI: 10.1016/j.aquatox.2009.06.010

[67] Seuront L. Behavioral fractality in marine copepods: Endogenous rhythms vs. exogenous stressors. Physica A: Statistical Mechanics and its Applications. 2011;**309**:250-256. DOI: 10.1016/j.physa.2010.09.025

[68] Garaventa F, Gambardella C, Di Fino A, Pittore M, Faimali M. Swimming speed alteration of *Artemia* sp. and *Brachionus plicatilis* as a sub-lethal behavioural end-point for ecotoxicological surveys. Ecotoxicology. 2010;**19**:512-519. DOI: 10.1007/s10646-010-0461-8

[69] Manfra L, Savorelli F, Di Lorenzo B, Libralato G, Comin S, Conti D, et al. Intercalibration of ecotoxicity testing protocols with *Artemia franciscana*. Ecological Indicators. 2015;**57**:41-47. DOI: 10.1016/j.ecolind.2015.04.021

[70] Pan RF. Experimental observation techniques and quantitative analysis methods used in zooplankton behavioral ecology. China: Ocean University of China; 2014. (Doctoral thesis)

[71] Libralato G. The case of *Artemia* spp. in nanoecotoxicology.

Marine Environmental Research. 2014;**101**:38-43. DOI: 10.1016/j.marenvres.2014.08.002

[72] Persoone G, Blaise C, Snell T, Janssen C, van Steertegem M. Cyst-based toxicity test: II-report on an international intercalibration exercise with three cost-effective toxkits. Zeitschrift für Angewandte Zoologie. 1993;**1**:17-34. DOI: 10.1016/S0140-6736(01)28882-5

[73] Little EE, Brewer SK. Neurobehavioral toxicity in fish. In: Schlenk D, Benson WH, editors. Target Organ Toxicity in Marine and Freshwater Teleosts New Perspectives: Toxicology and the Environment 2. London and New York: Taylor and Francis; 2001. pp. 139-174

[74] De Los Ríos P, Gajardo G. The brine shrimp *Artemia* (Crustacea; Anostraca): A model organism to evaluate management policies in aquatic resources. Revista Chilena de Historia Natural. 2004;77:3-4

Chapter 3

Phytoremediation of Hazardous Radioactive Wastes

Deepak Yadav and Pradeep Kumar

Abstract

Phytoremediation technology incorporates living plants for in situ remediation of contaminated soils, sediments, tailings and groundwater. These practices integrates the removal, or degradation of toxic wastes that is capable of cleaning up an area with low to moderate levels of contamination. Phytoremediation has been studied widely for metals, pesticides, solvents, explosives, crude oil, etc. These studies and research are advanced, especially in small-scale operations. Phytoremediation has been successfully tested to decontamination of radioactive sites. The chapter initiates with possible remediation methods used for radioactive wastes where we will discuss types and nature of radioisotope contamination. Then we discuss discusses the classifications of phytoremediation techniques to treat radioactive contaminated waste. Phytoremediation performance depends on numerous factors such as soil composition, level of toxicity, suitable plant species, etc. Conversely, phytoremediation prospects low cost, practical and ecologically viable approach for low-level radiation waste clean-up.

Keywords: phytoremediation, plants, radioactive pollutants, radioisotope, low-level radiation waste

1. Introduction

The term "Phytoremediation" derived from Greek and Latin words. The word "Phyto" from Greek "phutón" meaning plants, while the word "remediation" from Latin "remedium" meaning a remedy or cure or correct evil. Phytoremediation is a fairly new technology introduced in 1980s that use plants to clean/partly clean contaminated locations, or reduce the contaminants less harmful [1–4]. It is also named as green remediation, agro-remediation, botano-remediation, and vegetative remediation [4, 5]. Therefore, the word is relating the technology for the management of environmental problems using plants and their allied microorganisms.

Many observes, phytoremediation, especially when very toxic materials are in request (e.g., radioactive waste). Phytoremediation is a general term refers to a set of plant–contaminant interactions, and a precise application procedure involved in remediation of radioactively contaminated locations. Most of these practices involve applying information known for decades in agriculture and ecology to environmental problems. Basic information of phytoremediation comes from several research areas; containing ecotoxicology, plant ecophysiology, agriculture, toxicity and translocation of toxic radioactive isotopes.

Mitigation of the environmental pollutant challenges with excavating the contaminants, dispose them off somewhere else with a cost-effective and an environmentally friendly technique in waters and soils purification even in heavy metals [2, 3]. The conventional remediation techniques like chemical, thermal, physical and other treatment methods are costly, and may cause more contaminations to the environments [4].

The International Atomic Energy Agency (IAEA) has clearly defined radioactive waste as those materials that emits radioactive particles comprise intensities greater than the prescribed benign on national and international standards, and no additional use is expected [5–7]. For instance, hazardous radioactive waste is produced at each stage of the nuclear (uranium) fuel cycle with no nuclear waste facility. So, there is an immediate need of permanent storage facilities as well as repositories for the high-level nuclear wastes. Radioactive waste can be solid, liquid or gas from a diverse group of operations and activities (mining to nuclear power) and accidents poses health risks and has the potential to interrupt ecosystems.

Phytoremediation of radioactive waste is a method that uses plants to remove, transfer, or immobilize radionuclides present in the contaminated soil, water, or sludge, and it is a useful method for treating large-scale but low-level radionuclide pollution. Radioactive materials provide numerous applications (scientific, medical, agricultural, industrial and energy generation) and play a significant role in daily life in human society. Consequently, it is predictable that such diverse actions lead to radioactive waste generation. The nuclear accident at Chernobyl, Ukraine alone has been calculated to have increased the risk of cancer to humans by 0.1% [2, 8].

Radioactive uranium (U), caesium (Cs), strontium (Sr), and plutonium (Pu) are the main radioactive isotopes present in the environment as a consequence of nuclear activities, and are the radionuclides of most concern (for a list of radionuclides of environmental and health concern. Sometime the radioactive wastes have military applications, e.g., depleted uranium is used in weaponries, and the spent nuclear fuel (from reactors) comprise weapons-usable plutonium. Nuclear waste containing short radiation, generally of little concern as it fades quickly by natural radioactive decay. Conversely medium-level long-lived and high-level radioactive nuclear waste is more challenging and benign disposal of this waste is essential. Most of the nuclear waste produced in nuclear power plants, (half-life and effects of environmentally dangerous radioactive isotopes are listed in **Table 1**).

In specific, the prerequisite for the concern is for spent (used) radioactive fuel recently removed from nuclear reactors. Moreover, there is an alarming accumulation of radioactive material cast in glass or ceramics, shielded in stainless steel containers which is held in dry storage across the world [10, 11]. So, radioactive waste needs to be managed in a safe and must be remote from people till it remains dangerous.

Radionuclide	Half-life	Uses	Effects
Uranium (238)	4.5 billion years	Bombs, weapons, nuclear fuel	Mutations, cancer, birth defects
Plutonium (244)	80.8 million years	Explosives, mixed oxide fuel, power and heat sources. Example atomic bombings of Hiroshima and Nagasaki	Fire hazard, radioactivity and the heavy metal poison, radiation sickness, genetic damage, cancer, and death
Thorium (232)	14 billion years	Alloying agent, nuclear fuel	Carcinogenic
Radium (226)	1601 years	Luminous paints, dials of watches	Lymphoma, leukaemia, bone cancer

Table 1.
Phytoremediation of radioactive metals [9].

Radionuclides waste sources are transported in soil, sediments, or sludges can be reduced over and done with absorption and accumulation by the plant roots; adsorption onto roots; precipitation, or reduction in soil with root zone; or binding to humic (organic) matter by the process of humification. Before phytoremediation of the concerned radioactive waste, the appropriate natural plant should be wisely selected. The ways for selection the right plant species for phytoremediation of the radioactive waste are as follow:

1. Primarily, the features of radioactive waste should be examined.

2. Next, the plant class and its composition should be recorded.

3. Then, the concentration of a concerned radionuclide in the plant should be determined.

4. The plant biomass should be considered, and

5. Lastly, the concentration of a goal radionuclide in the remediated radioactive waste should be restrained.

Phytoremediation of radioactive waste is a worthwhile technique for treating large-scale, but low-level radionuclide waste. However, from the above mentioned criteria we can screen out the right plant types proficient to remediate the concerned radioactive waste. In the present chapter, significant features prompting the choice of natural plant to remediate radioactive waste. The concentration and features of radioactive waste, the plant type and plant structure, deposited area are detected, and the standards based on the phytoremediation factor (PF) have been anticipated for the selection of natural plant to phytoremediate radioactive waste.

2. Classification of radioactive wastes

Radioactive waste is distinct radioactive material for which no further use is foreseen in gaseous, liquid or solid form and controlled by a regulatory organization. According to international law governed by IAEA, spent nuclear fuel is not defined as wastes are well-defined by the accountable country. The wastes are categorized by the type and concentration of radioactive particles emitted (α, β and γ), energy and heat generation. The latest waste classification system for radioactive waste has been approved in universal standards established by the IAEA and are explained as follows [7]:

Exempt waste (EW): It comprises such a low concentration of radionuclides that create negligible radiological hazards and it can be excluded from nuclear regulatory control.

Very short lived waste (VSLW): These types of wastes are often treated to achieve volume reduction and/or conditioned, stored for decay over a limited period of few years, prior to disposal. These are disposed of as regular industrial waste and consequently cleared of regulatory control [12, 13]. Further, various safe and effective treatment routes are open, with chemical precipitation as well as incineration.

Very low level waste (VLLW): It does not require isolation and a high level of containment, and disposal is done in near-surface landfill. VLLW wastes are always cured to attain liquidity (volume) reduction and waste is immobilized prior to its disposal [13, 14]. Several safe and effective additional treatments are available, e.g., chemical precipitation and incineration.

Low level waste (LLW): It covers limited amounts of long-lived radionuclides with a very wide variety of radioactive waste. Waste that does not need shielding for handling or transportation, and isolation ages of a few 100 years. LLW may be slightly contaminated with radiation; for example, paper, glassware, tools and clothing. A wide range of disposal and storage alternatives are available, from simple to complex engineered facilities, e.g., landfills or incineration.

Intermediate level waste (ILW): ILW (reactor components, chemical residues, used metal fuel cladding) contains long-lived radionuclides alpha (α) emitters and isolation blocks. It does not need facility of heat dissipation during storage and disposal. ILW requires special handling and shielding of radioactivity. This waste is destined for disposal in deep geological repositories (the Waste Isolation Pilot Plant in USA).

High level waste (HLW): HLW covers high intensities of radiations that produce major amounts of heat by radioactive degeneration. It demands the design of removal in very deep, even geological layers, typically several hundred meters below the surface. The two primary categories are: (1) used fuel rods from nuclear plants and (2) waste from reprocessing the fuel rods. The waste contains both short-lived and long-lived high radiation nucleotides (half-lives of many thousands of years) which comprises high concentrations of radioactivity and requires cooling and special shielding, handling and storage.

3. Treatment methods

Conventional remediation techniques e.g., chemical, thermal and physical treatment methods are too costly, and may end of causing more contamination to the environment. Internationally acclaimed phytoremediation has an over 300-year old history of wastewater discharges, but the concept of using plants for the remediation of heavy metals and other pollutants was first reported in 1983 [15]. The concentration of a target element governs the degree to the widespread phytoremediation. Phytoremediation might be best suited for positions with the levels of radionuclide pollution which are only slightly advanced than the cleanup board levels because the subsequent sum of time for cleaning becomes reasonable (<10 years) and as probable plant toxicity effects are avoided [16].

Once the action is finished, an inorganic deposit remains that must be disposed of carefully, this residue has no fiscal significance. There are five varieties for positioning hazardous waste:

1. Hazardous wastes are dumped by force and under pressure by underground instillation bores (steel- and concrete-encased channels under earth crust).

2. Surface impoundment (engineered or natural depressions) can be recycled to treat, store, or dispose of hazardous waste, in pits or diked spaces.

3. Land-fills are discarding facilities where hazardous waste is located in, properly planned and lined landfills to prevent leakage.

4. Land treatment is a disposal process in which natural microbes in the soil break down (immobilize) the hazardous constituents.

5. Waste piles, non-flowing hazardous waste are used for provisional loading till it is moved to final removal and final disposal.

The hazardous waste disposed of on land, ~60% (underground injection wells), ~35% (surface impoundments), 5% in landfills, and <1% in waste piles/land application.

Radioactive waste control involves reducing radioactive residues, manage waste-packing carefully, safe storage and disposal along with protect sites of radioactivity origin clean. Underprivileged practices may lead to future complications. Therefore, selection of sites where radioactivity is to be managed safely is equally important other than technical expertise and investment, to result in safe and ecologically sound results. IAEA is endorsing recognition of some basic tenets by all countries for radioactive waste management which include:

i. Acquiring adequate level of human safety.

ii. Facility of a standard level of environment protection.

iii. Although predicting (i) and (ii), guarantee of insignificant properties past national boundaries.

iv. Tolerable impact on future groups, and

v. No unnecessary liability on future generations. There are other legal, control, generation, safety and management characteristics likewise.

The following decisions have been declared stress staid studies and technical assessments:

- Deep geological sources.
- Ocean dumping Seabed burial.
- Sub-seabed disposal.
- Subductive waste disposal method.
- Transforming radioactive waste to non-radioactive stable waste.
- Dispatching to the Sun.

4. Radioactive waste uptake phytoremediation mechanisms

Phytoremediation is well accepted in literature [17, 18], and favored due to its in-situ/ex-situ applicability. Further additional benefits comprise fairly easy to handle and apply, proficient extraction bioavailable shares of pollutants, adaptable to a range of organic and inorganic complexes and energy generation. While the use of plants as environmental rehabilitation agents has gained wide acceptability in multidisciplinary research fields. Bramley-Alves et al. [19] proposed that phytoremediation involves a multi-skill technique for example phyto-oxidation, volatilization, and microbial remediation to improve the efficacy of pollutants' control [19]. Furthermore, phytoremediation is favored to former chemical methods because it could be useful in locations contaminated by inorganic (e.g., heavy metals) contaminants and organic (e.g., pesticides, polycyclic aromatic hydrocarbons (PAH), and polychlorinated biphenyls) [20].

Over the previous years, several methods have been used to deal with the radioactive waste from contaminated sites. Though, these methods are costly and inefficient in their concert. The chemical methods generate large volumes of sludge and increase the cost of maintenance. Thermal methods are technically difficult and adversely affect the valuable component of soil by degrading it [21]. Two major procedures that are conventionally used to remediate the radioactive contaminated sites are: [22].

1. ***Ex-situ methods***: This method requires the removal of contaminated soil for treatment on/off site and then returning the treated soil to the site. The example of ex-situ methods are; soil leaching, solidification, immobilization, vitrification, heap leaching, ground disposal, sea disposal, incineration, and or destruction etc.

2. ***In-situ methods***: In this method excavation of contaminated location is not needed. The examples are; de-chlorination, bottom sealing, electromagnetic heating, etc.

Phytoremediation is a novel resolution that effectively and inexpensively extracts out the contaminants from the site and scrubs up the wasteland [23]. Phytoremediation makes use of green plants to clean up and treat radioactive contaminated sites for example soil, water and sediments. Plants have notable features that help them absorb contaminants into their systems with their endorsement capabilities such as translocation, bioaccumulation and contaminant degradation. Many plant species have been successful in efficiently accumulating the radionuclides in their stems and leaves and hence remediating the contaminated site [21]. This chapter evaluates some of the research that has been done on phytoremediation of radioactive metals and aims to discuss the potential of phytoremediation, highlight the general mechanisms of plant uptake, give a brief overview on radioactive metals (especially: Uranium-238, Thorium-232, Radium-226) uptake by plants, and report the advantages and limitations associated with this method.

5. Six main subgroups in phytoremediation

1. **Phytoextraction:** Plants degrading pollutants from the soil (tailings) and concentrating the contaminants in the harvestable portions of plants; i.e., in all organs of the plant—leaves, stems and roots [24, 25].

2. **Phytodegradation:** Plants removing pollutants by using hydrolytic enzymes and metabolites in plants; however, this method may be limited only to degradation of organic contaminants [26, 27].

3. **Phytostabilization:** Plants reducing mobility and bioavailability of pollutants in the soil either by immobilization and precipitation, or by preventing contaminant migration [28, 29].

4. **Phytovolatilization:** Volatilization of pollutants into the air directly or indirectly via plant uptake into tissues and organs, and then transformation of the products into volatile compounds [25, 30].

5. **Rhizofiltration:** Plant roots strongly absorbing, accumulating and/or precipitating contaminants from aqueous waste streams or soil water almost exclusively into the root system [31, 32].

6. **Rhizodegradation:** Enhancement of naturally occurring biodegradation and destruction of contaminants in the soil through mineralization and transformation of pollutants by plant roots and associated microbes [33–35].

6. Factors affecting the uptake mechanisms

There are several factors which can affect the uptake mechanisms of radioactive metals and are discussed as below:

6.1 Plant species

Plant species with superior remediation ability of the concerned radioactive waste are screened and carefully chosen. The success of phytoremediation technique depends upon the ability of the plant to accumulate [36].

6.2 Properties of medium

Factors such as temperature, moisture content, pH, organic matter affect the rate of uptake by plants [37].

6.3 The root zone

It can absorb contaminants and store or metabolize it inside plant tissue. An increase in root diameter and reduced root elongation as a response to less permeability of the dried soil [38].

6.4 Addition of chelating agents

The increase of the uptake by crops can be influenced by increasing the bioavailability of radionuclides through addition of biodegradable physiochemical factors such as chelating agents, and micronutrients [39].

In a stressed environment, the application of plants to remediate sites governed by mainly on the persistence capacity of the plant. All through phytoremediation, plants absorb pollutant from the soil, and mineralized it, thus preventing infection of groundwater and retaining system shield for human habitation. Efe and Elenwo reported that plants (e.g., *Axonopus compressus*) used as phytoremediation means, should have the capacity to adapt properly to the climatic condition and soil of the polluted sites, and retain high patience under stressed environments [40]. Several phytoremediation plant types have technologically advanced adaptive features for absorption, acceptance, transfer and degradation of pollutants for example heavy metals, crude oil, explosives, and radionuclides [41].

The efficiency of the process is also dependent on the soil properties, type of contaminants and its bioavailability. Plant roots usually serve as interlinks providing enormous surface area for the absorption and accumulation of essential growth nutrient along with contaminants [42]. In metal contaminated sites, characterization of eco-toxicity (e.g., oxidative stress) is mostly determined through the formation of free radicals [43]. Some of the advantages of phytoremediation include risk containment, extraction of valuable metals (phytomining) and increased soil fertility/quality.

Baker and Brooks [44] recommended that the metal hyperaccumulator must fulfill a standard that the concentration of an element stored in a plant can be higher than the soil [44]. Based on their classification, the transfer factor (TF) can be defined as the ratio of target element concentration in the plant to that in the tailings.

$$\text{Transfer factor} = (\text{target element concentration in the plant}) / (\text{target element concentration in the tailings}). \quad (1)$$

TF can be used as an index for the growth of a target element in the plant and its transfer from the tailings to the plant. If TF for a plant is greater than 1 and the amount of the target element collected in the plant is relatively small, the elimination competence of the plant for that target element can be further improved by a number of breeding practices, and can further implemented in phytoremediation [45].

Different TF values for the plants tissues may be resulted in part from metabolic rate differences between plant species and cultivations [46]. The factors for example the concentration of a radionuclide, pH, plant age, and ecotype may adjust the uptake and ratio of the content of the element present in the plant shoot to that in its root [47]. About 91 tissues of plant species had the TF values of <1, only 9 tissues of plant species had the TF values of more than 1. Overall, it was found that most of the plant species inspected had low experiences of removing U, Th, and ^{226}Ra from the stakeouts to the plant tissues. The results were friendly with the earlier research results [48–52].

In summary, phytoremediation of goal radionuclides from the followings largely depends primarily on three parameters with the radionuclide concentration in the plant, the plant biomass, and the target radionuclide concentration in the investigations. In order to assess the potential of a plant for phytoremediation more broadly, a novel coefficient was anticipated and named as phytoremediation factor [53]. This factor is the ratio of the total amount of a target radionuclide accumulated in the plant shoot to the concentration in the tailings at the site where the plant grows.

$$\begin{aligned} &\text{Phytoremediation factor (PF)} \\ &= (\text{target radionuclide concentration in the plant shoot}) / \\ &(\text{biomass of the plant shoot Target radionuclide concentration in the tailings}). \quad (2) \end{aligned}$$

In this formula, the shoot refers to the tissue above ground of the plant including the seed, leaf, and stalk. The PF can be used as an index for the capability of a plant to remove the target element from the tailings.

The results indicated that PF was agreeable with the plant removal capability. PF extends the conventional definition of hyperaccumulator, and it can easily be obtained. Although the concentration of a target radionuclide in a plant does not fulfill the criteria for a hyperaccumulator, if the plant has relatively high biomass, the plant may also be deliberated as the candidate for phytoremediation. Keeping in view the phytoremediation factor, *P. australis* and *M. cordata* were designated as the contenders for phytoremediation of uranium-contaminated soils [53, 54]. Azolla imbircata was selected as the candidate for phytoremediation of uranium-contaminated water [55, 56]. *P. australis* was selected as the candidate for phytoremediation of thorium-contaminated soils [53]. *P. multifida* was selected as the candidate for phytoremediation of ^{226}Ra-contaminated soils [54–56]. While PF offers a unique place for identification of a plant proficient in remediating the contaminated by the radioactive nuclides on a large scale, except the plant biomass per unit area. It is essential to consider further research should be executed to improvise this factor.

7. Phytomanagement

Waste disposal is dumping waste with no objective of retrieval. Waste management means the whole structure of operations starting with generation of waste and ending with disposal. The per capita use of electricity is correlated to the living standard of a country, whereas, the electricity generation by nuclear resources can be viewed as a least degree of radioactive waste that is produced and the allied scale of radioactive waste management of the country. On the gauge of electricity generation by nuclear fuel, India need to improve a lot. In 2000, India's stake of nuclear electricity generation compared to total electricity generation was 2.65% related to 75% of France which ranks first according to IAEA Report. Hence the magnitude of radioactive waste management in India could be miniscule compared to that in other countries.

As more power reactors come on stream and as weaponization takes profounder routes the needs of radioactive waste management increase. Radioactive waste management has been a crucial degree in the whole nuclear fuel cycle. Low and intermediate-level radioactive wastes rise from operations in reactors retained as sludge after chemical treatment and fuel reprocessing practices.

Solid radioactive waste is compressed, incinerated are subject to the nature of the waste. Underground drains in disposal facilities are applied for solid waste disposal under continuous surveillance and monitoring.

High efficiency particulate air (HEPA) filters are used to reduce air-borne radioactivity. From the last four decades radioactive waste management facilities have been set up at Trombay, Tarapore, Rawatbhata, Kalpakkam, Narora, Kakrapara, Hyderabad and Jaduguda, accompanied by the growth of nuclear power and fuel-reprocessing plants [57–63]. Numerous barrier methodology is monitored in solid waste handling in the next flow process are given below (eq. (3)):

$$\text{Source reduction} \rightarrow \text{Recycling} \rightarrow \text{Treatment} \rightarrow \text{Disposal.} \tag{3}$$

Flow process for management of waste reduction [57].

8. Conclusion and future directions

For the phytoremediation of radioactive waste, screening of the appropriate plant type is the utmost important. Diverse factors such as radioactive waste characteristics, the concentration of a target radionuclide in the radioactive waste, the biomass of the plant, the plant species and plants composition in the radioactive waste dumped area, the concentration of a target radionuclide in the plant, and should be examined thoroughly.

The PF concern the concentration of a goal element in a plant, the shoot biomass, and the concentration of the target element in the tailings or tailing (root) of the plant, was planned for the target element to specify the removal capability of the plant from the radioactive waste. Using the PF as the criteria, *P. australis*, *M. cordata*, and *Azolla imbricata* were selected as the contenders for phytoremediation of uranium-contaminated soil, *P. multifida* was particular as the aspirant for phytoremediation of ^{226}Ra-contaminated soil, and *P. australis* was designated as the contestant for phytoremediation of thorium-contaminated soil.

Further advances must be made in the application of environmental remediation to selectively eradicate materials, the concentrations of chemicals present in the contaminated water, have a higher resistance to changes in pH, greater stability for a longer period of time and cost effectiveness.

Sensors have been established for detecting gases, chemicals and volatile organic compounds (VOCs), and the detection and identification of radiations. Further growth is essential in the functional properties of nanomaterials to meet the requisite for trace detection and the treatment of pollutants in soil, water and air and important fundamental and mechanistic studies are required in order to fully explore their real potentials. The CNTs/metal oxide are promising constituents in ecological pollution management at a bigger prospective for practical applications.

Conflict of interest

The authors declare that there is no conflict of interests regarding the publication of this book chapter.

Author details

Deepak Yadav[1] and Pradeep Kumar[2]*

1 Department of Chemical Engineering, HBTU, Kanpur, India

2 Department of Chemical Engineering and Technology, IIT (BHU), Varanasi, India

*Address all correspondence to: pkumar.che@iitbhu.ac.in

References

[1] Barkay T, Schaefer J. Metal and radionuclide bioremediation: Issues, considerations and potentials. Current Opinion in Microbiology. 2001;**4**(3):318-323

[2] Dushenkov S. Trends in phytoremediation of radionuclides. Plant and Soil. 2003;**249**(1):167-175

[3] Ibeanusi VM, Grab DA, Jensen L, Ostrodka S. Radionuclide biological remediation resource guide. US Environmental Protection Agency, Region 5, Superfund Division; 2004

[4] Eapen S, Singh S, D'Souza SF. Phytoremediation of metals and radionuclides. In: Singh SN, Tripathi RD, editors. Environmental Bioremediation Technologies. Berlin Heidelberg: Springer-Verlag; 2007

[5] Olson PE, Fletcher JS. Ecological recovery of vegetation at a former industrial sludge basin and its implications to phytoremediation. Environmental Science and Pollution Research. 2000;**7**(4):195-204

[6] Pivetz BE. Ground Water Issue: Phytoremediation of Contaminated Soil and Ground Water at Hazardous Waste Sites. Ada, OK: National Risk Management Research Lab; 2001

[7] IAEA (International Atomic Energy Agency). Managing Radioactive Waste. Vienna: International Atomic Energy Agency; 2010. Available at: www.iaea.org/books/

[8] IAEA (International Atomic Energy Agency). Chernobyl's legacy: Health, environmental and socio-economic impacts. In: Cherbobyl Forum 2003-2005. Vienna: IAEA; 2005 Available at: www.iaea.org/books/

[9] Malhotra R, Agarwal S, Gupta P. Phytoremediation of radioactive metals. Journal of Civil and Environmental Engineering Technology. 2014;**1**:75-79

[10] Le Bars Y, Pescatore C. Shifting paradigms in managing radioactive waste. NEA News. 2004. 14-6

[11] Eskander S, Saleh H. Phytoremediation: An overview. In: Environmental Science and Engineering, Soil Pollution and Phytoremediation. 2017, 11. pp. 124-161

[12] De Filippis LF. Role of phytoremediation in radioactive waste treatment. In: Hakeem K, Sabir M., Ozturk M, Mermut A, editors. Soil Remediation and Plants: Prospects and Challenges. 1st ed. Academic Press; 2014. p. 207-254. DOI: 10.1016/B978-0-12-799937-1.00008-5.ch8

[13] Eskander SB, Saleh HM. Using Portland cement for encapsulation of *Epipremnum aureum* generated from phytoremediation process of liquid radioactive wastes. Arab Journal of Nuclear Sciences and Applications. 2010;**43**:83-92

[14] Dimitrescu I. Technology of underground storage of radioactive waste. Review Mineral Mining. 2010;**6**:15-19

[15] Blaylock M. Phytoremediation of Contaminated Soil and Water: Field Demonstration of Phytoremediation of Lead Contaminated Soils. Boca Raton, FL: Lewis Publishers; 2008

[16] Schnoor J. Phytoremediation of soil and groundwater. Prepared for the ground-water remediation technologies analysis center. Technology Evaluation Report TE-02-01. U.S. Department of Energy. Phytoremediation: natural attenuation that really works. TIE Quarterly. 1997;**6**(1)

[17] Njoku KL, Akinola MO, Oboh BO. Phytoremediation of crude oil contaminated soil: The effect of growth of *Glycine max* on the physico-chemistry and crude oil contents of soil. Nature and Science. 2009;**7**(10):79-87

[18] Njoku KL, Akinola MO, Anigbogu CC. Vermiremediation of soils contaminated with mixture of petroleum products using *Eisenia fetida*. Journal of Applied Sciences and Environmental Management. 2016;**20**(3):771-779

[19] Bramley-Alves J, Wasley J, King CK, Powell S, Robinson SA. Phytoremediation of hydrocarbon contaminants in subantarctic soils: An effective management option. Journal of Environmental Management. 2014;**142**:60-69

[20] Zhang Y, Liu J, Zhou Y, Gong T, Wang J, Ge Y. Enhanced phytoremediation of mixed heavy metal (mercury)–organic pollutants (trichloroethylene) with transgenic alfalfa co-expressing glutathione S-transferase and human P450 2E1. Journal of Hazardous Materials. 2013;**260**:1100-1107

[21] Pavel LV, Gavrilescu M. Overview of ex situ decontamination techniques for soil cleanup. Environmental Engineering and Management Journal (EEMJ). 2008;**7**(6)

[22] Ghosh M, Singh SP. A review on phytoremediation of heavy metals and utilization of it's by products. Asian Journal Energy Environment. 2005;**6**(4):18

[23] Salt DE, Blaylock M, Kumar NP, Dushenkov V, Ensley BD, Chet I, et al. Phytoremediation: A novel strategy for the removal of toxic metals from the environment using plants. Bio/Technology. 1995;**13**(5):468

[24] Kumar PBAN, Dushenkov V, Ensley BD, Chet I, Raskin I. Phytoremediation: A novel strategy for the removal of toxic metals from environment using plants. Biotechnology. 1995;**13**:1232-1238

[25] Banuelos GS, Ajwa HA, Mackey B, Wu L, Cook C, Akohoue S, et al. Evaluation of different plant species used for phytoremediation of high soil selenium. Journal of Environmental Quality. 1997;**26**(3):639-646

[26] Burken JG, Schnoor JL. Predictive relationships for uptake of organic contaminants by hybrid poplar trees. Environmental Science and Technology. 1998;**32**(21):3379-3385

[27] Schröder P, Navarro-Aviñó J, Azaizeh H, Goldhirsh AG, DiGregorio S, Komives T, et al. Using phytoremediation technologies to upgrade waste water treatment in Europe. Environmental Science and Pollution Research International. 2007;**14**(7):490-497

[28] Smith RA, Bradshaw AD. The use of metal tolerant plant populations for the reclamation of metalliferous wastes. Journal of Applied Ecology. 1979:595-612

[29] Vangronsveld J, Herzig R, Weyens N, Boulet J, Adriaensen K, Ruttens A, et al. Phytoremediation of contaminated soils and groundwater: Lessons from the field. Environmental Science and Pollution Research. 2009;**16**(7):765-794

[30] Burken JG, Schnoor JL. Distribution and volatilization of organic compounds following uptake by hybrid poplars. International Journal of Phytoremediation. 1999;**1**:139-151

[31] Dushenkov V, Kumar PN, Motto H, Raskin I. Rhizofiltration: The use of plants to remove heavy metals from aqueous streams. Environmental Science and Technology. 1995;**29**(5):1239-1245

[32] Vara Prasad MN, de Oliveira Freitas HM. Metal hyperaccumulation in plants: Biodiversity prospecting for phytoremediation technology. Electronic Journal of Biotechnology. 2003;**6**(3):285-321

[33] Zhou Q, Cai Z, Zhang Z, Liu W. Ecological remediation of hydrocarbon contaminated soils with Weed Plant. Journal Resour Ecology. 2011;**2**(2):97-105. DOI: 10.3969/j.issn.1674-764x.2011.02.001

[34] Rugh CL. Mercury detoxification with transgenic plants and other biotechnological breakthroughs for phytoremediation. In Vitro Cellular and Developmental Biology: Plant. 2001;**37**(3):321

[35] Sors TG, Ellis DR, Salt DE. Selenium uptake, translocation, assimilation and metabolic fate in plants. Photosynthesis Research. 2005;**86**(3):373-389

[36] Fathi RA, Godbold DL, Al-Salih HS, Jones D. Potential of phytoremediation to clean up uranium-contaminated soil with acacia species. Journal of Environment and Earth Science. 2014;**4**(4):81-91

[37] Rodriguez L, Lopez-Bellido FJ, Carnicer A, Recreo F, Tallos A, Monteagudo JM. Mercury recovery from soils by phytoremediation. Environmental Chemistry. 2005:197-204

[38] Zhu YG, Smolders E. Plant uptake of radiocaesium: A review of mechanisms, regulation and application. Journal of Experimental Botany. 2000;**51**(351):1635-1645

[39] Ehlken S, Kirchner G. Environmental processes affecting plant root uptake of radioactive trace elements and variability of transfer factor data: A review. Journal of Environmental Radioactivity. 2002;**58**(2-3):97-112

[40] Efe SI, Elenwo EI. Phytoremediation of crude oil contaminated soil with *Axonopus compressus* in the Niger Delta region of Nigeria. Natural Resources. 2014;**5**(02):59

[41] Sinha RK, Valani D, Sinha S, Singh S, Herat S. Bioremediation of contaminated sites: A low-cost nature's biotechnology for environmental clean up by versatile microbes, plants and earthworms. Solid Waste Management and Environmental Remediation. 2009 978-1

[42] Raskin I, Ensley BD. Recent developments for in situ treatment of metal contaminated soils. In: Phytoremediation of Toxic Metals: Using Plants to Clean up the Environment. New York: John Wiley & Sons Inc.; 2000

[43] Arora M, Kiran B, Rani S, Rani A, Kaur B, Mittal N. Heavy metal accumulation in vegetables irrigated with water from different sources. Food Chemistry. 2008;**111**(4):811-815

[44] Baker AJ, Brooks R. Terrestrial higher plants which hyperaccumulate metallic elements. A review of their distribution, ecology and phytochemistry. Biorecovery. 1989;**1**(2):81-126

[45] Whicker FW, Hinton TG, Orlandini KA, Clark SB. Uptake of natural and anthropogenic actinides in vegetable crops grown on a contaminated lake bed. Journal of Environmental Radioactivity. 1999;**45**(1):1-2

[46] Chen SB, Zhu YG, Hu QH. Soil to plant transfer of ^{238}U, ^{226}Ra and ^{232}Th on a uranium mining-impacted soil from southeastern China. Journal of Environmental Radioactivity. 2005;**82**(2):223-236

[47] Tu C, Ma LQ, Bondada B. Arsenic accumulation in the hyperaccumulator Chinese brake and its utilization

potential for phytoremediation. Journal of Environmental Quality. 2002;**31**(5):1671-1675

[48] Soudek P, Petřík P, Vágner M, Tykva R, Plojhar V, Petrová Š, et al. Botanical survey and screening of plant species which accumulate ^{226}Ra from contaminated soil of uranium waste depot. European Journal of Soil Biology. 2007;**43**(4):251-261

[49] Soudek P, Petrová Š, Benešová D, Tykva R, Vaňková R, Vaněk T. Comparison of ^{226}Ra nuclide from soil by three woody species *Betula pendula*, *Sambucus nigra* and *Alnus glutinosa* during the vegetation period. Journal of Environmental Radioactivity. 2007;**97**(1):76-82

[50] Soudek P, Petrová Š, Benešová D, Kotyza J, Vagner M, Vaňková R, et al. Study of soil–plant transfer of ^{226}Ra under greenhouse conditions. Journal of Environmental Radioactivity. 2010;**101**(6):446-450

[51] Soudek P, Petrová Š, Benešová D, Dvořáková M, Vaněk T. Uranium uptake by hydroponically cultivated crop plants. Journal of Environmental Radioactivity. 2011;**102**(6):598-604

[52] Lauria DC, Ribeiro FC, Conti CC, Loureiro FA. Radium and uranium levels in vegetables grown using different farming management systems. Journal of Environmental Radioactivity. 2009;**100**(2):176-183

[53] Li GY, Hu N, Ding DX, Zheng JF, Liu YL, Wang YD, et al. Screening of plant species for phytoremediation of uranium, thorium, barium, nickel, strontium and lead contaminated soils from a uranium mill tailings repository in South China. Bulletin of Environmental Contamination and Toxicology. 2011;**86**(6):646-652

[54] Hu N, Ding D, Li G. Natural plant selection for radioactive waste remediation. In: Radionuclide Contamination and Remediation through Plants. Cham: Springer; 2014. pp. 33-53

[55] Hu N, Ding D, Li G, Wang Y, Li L, Zheng J. Uranium removal from water by five aquatic plants. In: Progress Report on Nuclear Science and Technology in China (Vol. 2). Proceedings of Academic Annual Meeting of China Nuclear Society in 2011, No. 2-Uranium Mining and Metallurgy Sub-Volume. 2012

[56] Hu N, Ding D, Li G, Zheng J, Li L, Zhao W, et al. Vegetation composition and ^{226}Ra uptake by native plant species at a uranium mill tailings impoundment in South China. Journal of Environmental Radioactivity. 2014;**129**:100-106

[57] Rao KR. Radioactive waste: The problem and its management. Current Science. 2001;**81**(12):1534-1546

[58] Saleh HM. Water hyacinth for phytoremediation of radioactive wastes simulate contaminated with cesium and cobalt radionuclides. Nuclear Engineering and Design. 2012;**242**:425-432

[59] Saleh HM. Stability of cemented dried water hyacinth used for biosorption of radionuclides under various circumstances. Journal of Nuclear Materials. 2014;**446**(1-3):124-133

[60] Saleh HM. Biological remediation of hazardous pollutants using water hyacinth—A review. Journal of Biotechnology Research. 2016;**2**(11):80-91

[61] Saleh HM, Bayoumi TA, Mahmoud HH, Aglan RF. Uptake of cesium and cobalt radionuclides from simulated radioactive wastewater by *Ludwigia stolonifera* aquatic plant. Nuclear Engineering and Design. 2017;**315**:194-199

[62] Bayoumi TA, Saleh HM. Characterization of biological waste stabilized by cement during immersion in aqueous media to develop disposal strategies for phytomediated radioactive waste. Progress in Nuclear Energy. 2018;**107**:83-89

[63] Saleh HM, Aglan RF, Mahmoud HH. *Ludwigia stolonifera* for remediation of toxic metals from simulated wastewater. Chemistry and Ecology. 2019;**35**(2):164-178

Section 3

Recycling and Disposal of Electronic Waste

Chapter 4

Electronic Waste Recycling and Disposal: An Overview

Cristina A. Lucier and Brian J. Gareau

Abstract

Electronic waste, or e-waste, is said to be the fastest growing stream of hazardous waste in the world. E-waste is comprised of a variety of inputs including hazardous materials, potentially valuable and recyclable materials, and other inputs. E-waste follows a range of pathways after disposal, including formal and informal recycling, storage, and dumping, in both developed and less-developed country contexts. Globally, the handling and regulation of e-waste as both a hazardous waste stream and as a source of secondary raw materials has undergone significant changes in the past decade. A growing number of countries have adopted extended producer responsibility laws, which mandate electronics manufacturers to pay for proper recycling and disposal of electronics. The e-waste recycling industry is becoming more formalized as the potential to recover valuable materials has increased, but a range of recent studies have shown that e-waste recycling continues to carry a range of occupational health and environmental risks.

Keywords: e-waste, waste electrical and electronic equipment, extended producer responsibility, Basel Convention

1. Introduction

Electronic waste, sometimes referred to as e-waste or waste electrical and electronic equipment (WEEE), is a highly varied stream of hazardous waste. This waste stream is comprised of any electronic items that a consumer or business intends to dispose of, or is no longer useful for its original purpose. E-waste has generated a considerable amount of public and political interest due to a confluence of factors, including: the exponential rise in the generation of e-waste, the potential value of recycling the waste in order to recover precious metals and other elements, and the environmental and human health risks associated with improperly storing, disposing of, and recycling e-waste. Some of the major responses to the rising generation of e-waste (and growing demand for secondary raw materials that it contains) have included the development of producer "take-back" legislation, technological innovations in recycling processes, and the formation of partnerships to facilitate the transfer of e-waste between the informal and formal recycling sectors [1].

E-waste is an incredibly complex waste stream, as it encompasses a wide range of items and the exact composition of many electronic components are considered to be trade secrets, meaning they are the confidential information of the manufacturer. Generally speaking, "modern electronics can contain up to 60 different elements; many are valuable, some are hazardous and some are both. The most complex mix of substances is usually present in the printed wiring boards (PWBs)" [2].

To use a specific example, the material content of a mobile phone includes "over 40 elements in the periodic table including base metals like copper (Cu) and tin (Sn), special metals such as cobalt (Co), indium (In) and antimony (Sb), and precious metals including silver (Ag), gold (Au), and palladium (Pd)" [2].

Electronics that had been used in industrial or business applications, such as medical equipment, have been recycled in the formal recycling industry for more than 40 years. These large items have frequently been exported within industrialized countries in the OECD to specialized facilities where they are processed for the purpose of extracting secondary raw materials. Consumer electronic waste from smaller items such as cell phones and televisions have not historically been profitable to recycle in countries with higher labor costs, since the quantity of recoverable valuable materials is relatively low. Hence, these items have typically either been disposed of, stored in consumers' homes, or exported (often illegally) to less developed countries such as China, India, Ghana and Nigeria, where they are recycled by informal recyclers using low-tech methods such as manual dismantling, open burning and acid leaching in order to recover gold, copper and other valuable metals. These methods generate subsistence livelihoods for workers but also result in significant hazards to human health and the environment as a result of the toxic materials that are also embedded in consumer electronics. This chapter will explore these conventional recycling efforts and the ways in which they are evolving alongside global economic developments and the introduction of new recycling processes and technologies.

Generally speaking, the e-waste recycling process consists of five basic stages: collection, toxics removal, preprocessing, end processing and disposal [3]. There are wide degrees of variation in how these stages are managed worldwide. For much of the global waste stream, e-waste may be collected informally via "waste pickers" or more formally through voluntary or mandatory producer "take-back" programs. In terms of consumer electronics, regions where e-waste is picked up by informal collectors have historically achieved significantly higher recycling rates than those where waste is dropped off through formal channels [4]. After reaching the recycling site, dangerous components that require special treatment (e.g., batteries, Freon) are removed. The units are then separated into more homogenous groups based on material. This can be done manually, mechanically or a combination of both. Manual dismantling involves tools such as screwdrivers, hammers and labeled containers, while mechanical dismantling may involve conveyor belts, giant shredders and magnets [5].

Following the separation and dismantling phases, more homogenous groups of material (e.g., gold, copper, plastic, circuit boards) are then treated through a refining process: this can be accomplished chemically, with heat, or with metallurgical processes. This stage can be as high-tech as a giant smelter in Antwerp, Belgium or as low-tech as acid stripping in a backyard in Guiyu, China. Research has uncovered how sites will often compete for the waste by offering low-cost strategies, sometimes described as a "race to the bottom" process of increasingly lower standards and environmental protection [6]. Finally, all of the components that cannot be sold or used as secondary raw materials are disposed of through means such as incineration or landfill.

The level of efficiency achieved through e-waste recycling depends upon the process that is followed, especially in the separation and dismantling phases. In dismantling electronics, manualized options are often much more effective than mechanized processes in gaining access to the best quality secondary raw materials. Mechanized take-back programs such as those in the E.U. do not even come close to the efficacy of the labor-intensive e-waste collection rates found in many African countries [4, 7]. Manual dismantling is also preferable to machine shredding, which damages and does not completely separate individual materials. For example, while 90% of the gold in discarded mobile phones can be recovered when manually dismantled, only 26% is recovered through mechanical shredding [8]. However, these more labor-intensive options are not cost effective unless labor costs are extremely low [3].

2. Secondary raw materials recovered in electronic waste recycling

E-waste contains components that have historically been valuable in significant quantities, when the dismantling costs have been low enough [9]. Some of the applications and quantities extracted for different "important" or valuable elements within electronic devices are represented in **Table 1**.

In addition to these metals, there is also another subset of elements—known as rare earth elements—which are crucial to the functioning of the newest electronics, particularly those with LED lighting and touch screen technologies. Rare earth elements are available in abundant quantities globally, but the process of their extraction can create widespread environmental problems, including radioactive contamination [10]. **Table 2** provides a list of the rare elements that are used in various electronics. It is worth noting that the actual quantity of these elements used is relatively small, but that their properties are closely linked to the performance level of these technologies [11]. Rare earths play a particularly decisive role in the high performance functioning of magnets. The information provided in **Table 2** has been adapted from information derived from the U.S. Department of Energy, a report commissioned for the U.S. Interior Department and the U.S. Geological Survey, as well as industry trade publications [12–14]. Those rare earths considered to be of the highest potential resale value (and the highest risk for supply shortages) are neodymium (Nd), europium (Eu), dysprosium (Dy), terbium (Tb) and yttrium (Y) [12, 14].

Recent technological developments, including improvements to the mechanization process as well as pilot projects that combine low-tech and mechanized

Element	Main applications	Total tons/year [2006]
Silver	Contacts, switches, solders	6000
Gold	Bonding wire, contacts, integrated circuits	300
Palladium	Multilayer capacitors, connectors	33
Platinum	Hard disk thermocouple, fuel cell	13
Ruthenium	Hard disk, plasma displays	27
Copper	Cable, wire connector	4,500,000
Tin	Solders	90,000
Antimony	Flame retardant; CRT glass	65,000
Cobalt	Rechargeable batteries	11,000
Bismuth	Solders, capacitor	900
Selenium	Electro-optic copier, solar cell	240
Indium	LCD glass, solder, semiconductor	380

Source: [2].

Table 1.
A sample of valuable elements in electronic wastes.

Technology	Rare earths used
Electric and hybrid cars (NiMH battery)	Neodymium, praseodymium, dysprosium, terbium
Computers (magnets in hard disk drive)	Neodymium, praseodymium, dysprosium, terbium
Flat panel screens (glass coating to produce colors and brightness)	Yttrium, europium, terbium, gadolinium, praseodymium, cerium
MRI machines (magnets)	Neodymium, praseodymium, dysprosium, terbium, yttrium, europium
Smart phones (magnets and speakers)	Neodymium, praseodymium, dysprosium, terbium, yttrium, europium
Other uses (including "chemicals, military weapons and delivery systems, and satellite systems" ([13], p. 12)	Cerium, lanthanum, yttrium, neodymium, praseodymium, samarium, gadolinium

Sources: [12–14].

Table 2.
Common uses of rare earth elements in electronic devices.

methods, have been targeted to make e-waste recycling more profitable. Improvements to the mechanization process are fairly straightforward. On the one hand, revisions to shredding and sorting machines have improved the consistency and quality of the materials that are gathered at the preprocessing stage. In addition to this, newly mechanized methods are being developed to extract additional streams of secondary raw materials that were not previously recoverable. The major developments in this arena have been the invention of ways to extract various rare earth elements from electronics. State of the art facilities in Japan and France that can extract rare earths have recently become operational [15, 16]. Continued investment in technologies to recycle rare earths is seen as a strategic priority of industrialized countries, as these materials are essential for technologies related to communications, defense, and other state objectives, yet most mining for these materials takes place in China, a global power that has recently imposed quotas on the quantities that it is willing to sell for export [12, 17–19]. Concerns over the security and stability of the supply of rare earths have driven the development of new mechanized technologies to recover these materials from a wide range of e-waste inputs. Cost effective technologies for recovering secondary neodymium, dysprosium and praseodymium from e-waste are being further developed by U.S.-based recyclers and research institutes [14]. Whether they are sited within the E.U., the U.S., or Japan, these newly operational recycling facilities will require a large quantity of e-waste inputs in order to be profitable. This challenge involves diverting a significant portion of e-waste from landfills, and from the informal recycling industry in less-developed countries.

3. The role of extended producer responsibility in e-waste recycling

Estimates of how much electronic waste is generated globally within a given year vary widely [3, 20]. These estimates are based on the quantity and volume of various electronic items that are purchased in a given year, with consideration to the anticipated life expectancy of that particular item [21]. Surveys of recyclers on the volume of electronics collected can also be factored in, but it is important to note that a significant portion of consumer e-waste is either stored in consumers' homes or is mixed in with regular household waste and disposed of into landfills [22].

In some instances, data on the amount of e-waste collected for recycling is available, such as in those regions that mandate producers to "take-back" consumers' unwanted electronics. Such mandates originated as part of the concept of extended producer responsibility (EPR), which holds that the manufacturers of products with hazardous components should bear the logistic and financial burden of recycling or disposing of their products in an environmentally responsible way [23].

While EPR legislation was initially opposed by manufacturers, the increased interest in the strategic importance and potential profitability of the secondary raw materials contained in e-waste (particularly the rare earth elements) has contributed to growing support for such legislation. Variations of EPR "takeback" laws have been put into effect across the globe, including in a number of U.S. states, across the European Union, and across many countries in Asia, Africa and Latin America [24]. These laws signal a potential shift away from electronics recycling taking place primarily in the informal sector, and towards the growth of the formalized e-waste recycling industry.

With a growing number of EPR laws mandating manufacturers to take extended responsibility for the environmentally sound recycling of their products, there has been an increase in the number of pilot projects and public-private partnerships to collect and recycle electronics in ways that are efficient and cost-effective [25]. Some of these projects entail the transport of e-wastes across national boundaries, and have fallen under the purview of the Basel Convention on the Control of the Transboundary Movements of Hazardous Wastes and Their Disposal (the Basel Convention). Over the years, The Basel Convention has convened technical working groups and conducted pilot projects, both of which have resulted in the development of technical guidelines for the handling and management of e-waste [26–28].

Under the purview of the United Nations University, additional pilot projects are being developed to facilitate a globalized e-waste recycling chain that involves labor-intensive dismantling and preprocessing in countries with lower labor costs (e.g., China, India, African countries), and high-tech end-processing in countries with more modern facilities (e.g., the EU countries) [3, 25]. Major recycling corporations, electronics manufacturers, and government officials believe that such partnerships will insure a higher volume of input for large, high-tech smelters and provide access to the secondary raw materials that were previously "dumped" or otherwise retained within the global South countries where informal recycling currently takes place [2, 3].

4. Environmental and human health hazards of electronic waste recycling

The extent to which many of the other materials found in electronics are hazardous to human health and the environment is increasingly well-known. Electronics often contain toxic elements such as lead (Pb), cadmium (Cd), polychlorinated biphenyls (PCBs), polybrominated biphenyls (PBBs) and mercury (Hg) as well as other toxic components such as PVC and brominated flame retardants (BFRs) [29]. **Table 3** presents a list of some of the known hazardous components found in the typical desktop computer (with CRT monitor). This table is an adaptation of material presented by the Silicon Valley Toxics Coalition [30] in their report "Poison PC's and Toxic TV's" and toxicity data from Ceballos et al. [31].

Many of the health effects outlined in **Table 3** have been documented in the town of Guiyu, China, where perhaps the greatest portion of the U.S.'s e-waste exports have been deposited historically. Here, almost 80% of children have respiratory problems, and they have an especially high risk of lead poisoning [32]. Neurological, respiratory,

Element	Main applications	Weight (per 60 lb)	Dangers
Lead	Metal joining, radiation shield/CRT, printed wiring board	3.8	Human effects: neurological, blood, kidney damage. Brain damage and poisoning/death for children Accumulates in the environment
Mercury	Batteries, switches/ housing, printed wiring board	<0.1	Human effects: long term brain damage and other neurological effects Concentrates through the food chain.
Cadmium	Battery, blue/green phosphor emitter/housing, printed wiring board, CRT	<0.01	Human effects: acute and chronic damage to the kidneys
Plastics (including those containing PVC and BFRs)	Casing, cable coating	22.99	PVC effects: developmental toxin, reproductive toxin, endocrine disruptor. Carcinogenic when burned due to production of dioxins BFR effects: endocrine disruptor, neurotoxin, carcinogen (in humans and animals)
Sources: [30, 31].			

Table 3.
A sample of hazardous elements in an older-model PC.

digestive and bone problems are not uncommon among the workers and their families [32]. In addition to these toxicological threats, which include long-term implications for both health and the environment, the threats posed by the recycling of e-waste are even greater when certain recycling methods are employed [33–36]. For example, the informal recycling practice of burning plastic cables to retrieve the copper inside releases dioxins in the air via the burning PVC in the plastic. In more sophisticated operations (most typically found in Asian countries), a process of leaching printed circuit boards with acids (including nitric acid and hydrochloric acid) in order to maximize the amount of gold recovered can cause burns, respiratory and circulatory problems, pulmonary edema and death [2, 4, 36]. The acid stripping leaves behind a toxic residue that oftentimes is disposed of in waterways where it can acidify water and destroy wildlife and vegetation [36–38]. Heavy metal dust can travel to more populous areas and contaminate food supplies and greater populations [39, 40].

There have also been studies on the occupational health and environmental risks associated with high-tech e-waste recycling, with experts noting that much more research needs to be done in this area in order to gain a more accurate assessment of these risks [41–45]. Many of these studies are based on or informed by field research and experiments that measure concentrations of toxic chemicals in the workers, air, and environment around high-tech recycling facilities. There are indications in these studies that technologies such as the introduction of face masks and improved ventilation do decrease occupational exposures to a number of heavy metals and other hazardous chemicals.

In the U.S., a survey of 276 electronic waste recycling facilities was recently completed by the U.S. National Institutes of Occupational Safety and Health (NIOSH) [31]. It is especially relevant to note that this report finds that "most" of the responding facilities rely on manual dismantling, similar to the approach being applied in new and pilot facilities in less developed countries. It is also worth noting that a majority of the responding facilities were certified as environmentally sound either through the industry standard RiOS, the EPA standard R2 Solutions, or the activist standard e-Stewards. Hence, these facilities are likely to represent the "best case scenario" in electronics recycling. Overall, NIOSH concluded that "e-scrap recycling has the potential for a wide variety of occupational exposures particularly because of the use of manual processes" [31]. One of the primary concerns listed in the report is the potential for exposure to "metal dust" during the process of manual dismantling [31]. Specifically, the report notes that it is unclear whether most facilities have installed proper filtration systems in order to remove metal dust from the air (since the majority of the facilities circulate air within the production area or rely on "natural ventilation"). The report also notes the use of compressed air for cleaning which can heighten exposure to metal dusts. While the initial NIOSH report says that acute exposure to heavy metals such as lead is unlikely, the report notes that "chronic lead poisoning, which is more likely at current occupational exposure levels, may not have symptoms or they may have nonspecific symptoms that may not be recognized as being associated with lead exposure" [31].

Following the publication of the NIOSH report, additional studies of on-site occupational exposures in formal e-waste recycling facilities have been completed. Researchers reviewed 37 studies of the occupational hazards associated with formal e-waste recycling and concluded that, despite clear improvements to worker and environmental health when compared with informal recycling, "formal e-recycling workers and their families may experience unhealthful exposures to metals" [46]. The authors recommend further research "to reduce chemical exposures from formal e-waste recycling," along with the development of electronics components that are easier to safely disassemble, along with reducing the use of hazardous components in the manufacture of electronics [46]. With e-waste now considered to be the fastest growing stream of hazardous waste in the world, there is an urgent imperative to implement solutions to reduce the risks associated with e-waste recycling [47].

5. Conclusion

The design, production, sale and use of electronics takes place at the global scale. These initial stages in the life cycle of electronics pose a series of hazards to human health and the environment. Similarly, the disposal and recycling of electronics routinely entails the movement of hazardous materials across national borders. Growing government and industry interest in the recovery of secondary raw materials, such as the rare earth elements from e-waste, is leading towards an increase in the: development of strategies to increase recycling rates (which currently stand at approximately 20% globally), as well as in the development of formal, mechanized processes for recycling e-waste at the end-processing stage. In some cases, this has entailed the development of enabling legislation such as EPR "take-back" laws, and in other cases this has led to pilot projects that promote partnerships between recyclers in the formal and informal sectors. While there are many additional steps that can be taken in order to ensure that recipients of waste are adequately prepared to manage and recycle them in an environmentally sound manner, progress has been made. As these developments unfold, regulation and oversight will play a decisive role in mitigating the myriad risks to human health and the environment that can result from e-waste recycling.

Acknowledgements

This chapter was partially funded by the Boston College Open Access Publishing Fund.

Conflict of interest

The authors declare no conflict of interest.

Author details

Cristina A. Lucier[1] and Brian J. Gareau[2*]

1 West Palm Beach, FL, USA

2 Boston College, Chestnut Hill, MA, USA

*Address all correspondence to: gareau@bc.edu

References

[1] Lucier CA, Gareau BJ. Obstacles to preserving precaution and equity in global hazardous waste regulation: An analysis of contested knowledge in the basel convention. International Environmental Agreements: Politics, Law, and Economics. 2016;**16**:493-508. DOI: 10.1007/s10784-014-9261-6

[2] Solving the E-Waste Problem. Recycling—From E-Waste to Resources: Sustainable Innovation and Technology Transfer Industrial Sector Studies [Internet]. 2009. Available from: http://www.unep.fr/shared/publications/pdf/DTIx1192xPA-Recycling%20from%20ewaste%20to%20Resources.pdf

[3] Wang F, Huisman J, Meskers C, Schleup M, Stevels A, Hageluken C. The best of 2 worlds philosophy: Developing local dismantling and global infrastructure network for sustainable E-waste treatment in emerging economies. Waste Management. 2012;**32**(11):2134-2146. DOI: 10.1016/j.wasman.2012.03.029

[4] Secretariat of the Basel Convention. Where are WEEE in Africa [Internet]. Available from: http://www.basel.int/Portals/4/Basel%20Convention/docs/pub/WhereAreWeeInAfrica_ExecSummary_en.pdf

[5] Blacksmith Institute. In: The Top Ten Toxic Threats: Cleanup, Progress and Ongoing Challenges [Internet]. 2013. Available from: www.blacksmithinstitue.org

[6] Lucier CA, Gareau BJ. From waste to resources? Interrogating 'race to the bottom' in the global environmental governance of the hazardous waste trade. Journal of World-Systems Research. 2015;**21**(2):495-520

[7] Manhart A. E-Waste Africa Project: Impacts of current recycling practices and recommendations for collection and recycling. In: Pan African Forum on E-Waste; 14-16 March 2012; Nairobi, Kenya. 2012

[8] Rotter VS, Chancer P. Recycling of critical resources—upgrade introduction. In: Electronics Goes Green 2012+ Conference; 9-12 September 2012; Berlin, Germany. 2012. pp. 1-6

[9] Manomaivibool P, Lindhqvist T, Tojo N. Electrical and electronic equipment in India. Resources, Conservation and Recycling. 2009;**53**(3):136-144. DOI: 10.1016/j.resconrec.2008.10.003

[10] Schmitz O, Gradael T. The Consumption Conundrum: Driving the Destruction Abroad [Internet]. 2010. Available from: https://e360.yale.edu/features/the_consumption_conundrum_driving_the_destruction_abroad

[11] Gradael TE, Harper EM, Nassar NT, Reck BK. On the materials basis of modern society. Proceedings of the National Academy of Sciences. 2015;**112**(20):6295-67300. DOI: 10.1073/pnas.131275211

[12] DOE (US Dept. of Energy). Critical Materials Strategy [Internet]. 2010. Available from: http://www.doe.gov/sites/prod/files/edg/news/documents/criticalmaterialsstrategy.pdf

[13] Goonan TG. Rare Earth Elements—End Use and Recyclability: U.S. Department of the Interior, U.S. Geological Survey Scientific Investigations Report 2011-5094 [Internet]. 2011. Available from: https://pubs.usgs.gov/sir/2011/5094/

[14] Harler K. Rare Opportunity to Recycle Rare Earths [Internet]. 2018. Available from: https://www.recyclingtoday.com/article/rare-earth-metals-recycling/

[15] Akahori T, Hiroshige Y. Economical evaluation of recycling system for

rare-earth magnets. Electronics Goes Green 2012+ Conference; 9-12 September 2012; Berlin, Germany. 2012. pp. 1-5

[16] Recycling Today. Solvay Opens Rare Earths Metals Recycling Plants in France [Internet]. 2012. Available from: https://www.recyclingtoday.com/article/solvay-rare-earth-recycling-plants-france/

[17] Jowitt SM, Werner T, Weng Z, Mudd G. Recycling of the rare earth elements. Current Opinion in Green and Sustainable Chemistry. 2018;**13**:1-7

[18] WTO. China—Measures Related to the Exportation of Rare Earths, Tungsten and Molybdenum: Dispute Settlement: Dispute DS431 [Internet]. 2014. Available from: https://www.wto.org/english/tratop_e/dispu_e/cases_e/ds431_e.htm

[19] EC (European Commission). Tackling the Challenges in Commodity Markets and On Raw Materials [Internet]. 2011. Available from: http://eurlex.europa.eu/LexUriServ/LexUriServ.do?uri=COM:2011:0025:FIN:EN:PDF

[20] Terazono A, Murakami S, Abe N, Inanc B, Moriguchi Y, Sakai S, et al. Current status and research on e-waste issues in Asia. Journal of Material Cycles and Waste Management. 2006;**8**:1-12. DOI: 10.1007/s10163-005-0147-0

[21] EMPA. International E-waste generation [Internet]. 2005. Available from: http://www.ewaste.ch/facts_and_figures/statistical/quantities

[22] EPA (US Environmental Protection Agency) 'E-waste webpage'. Available from: http://www.epa.gov/osw/conserve/materials/ecycling/manage.htm

[23] Organization for Economic Cooperation and Development (OECD). EPR Policies and Product Design: Economic Theory and Selected Case Studies: OECD Report NV/EPOC/WGWPR(2005)9/FINAL [Internet]. 2005. Available from: http://www.oecd.org/fr/env/dechets/factsheetextendedproducerresponsibility.htm

[24] International Telecommunications Union. Chapter 10: Regional E-Waste Status and Trends [Internet]. 2017. Available from: https://www.itu.int/en/ITU-D/Climate-Change/Pages/Global-E-waste-Monitor-2017.aspx

[25] Solving the E-Waste Problem (StEP). StEP Projects [Internet]. Available from: https://step-initiative.org/projects-59.html

[26] Basel Convention Secretariat. Environmentally Sound Management Toolkit [Internet]. Available from: http://www.basel.int/Implementation/CountryLedInitiative/EnvironmentallySoundManagement/ESMToolkit/Overview/tabid/5839/Default.aspx

[27] Basel Convention Secretariat. Pilot Projects [Internet]. Available from: http://www.basel.int/Implementation/CountryLedInitiative/EnvironmentallySoundManagement/ESMToolkit/Pilotprojects/tabid/5846/Default.aspx

[28] Partnership for Action on Computing Equipment. Pilot Projects [Internet]. Available from: http://www.basel.int/Implementation/TechnicalAssistance/Partnerships/PACE/Pilotprojects/tabid/5381/Default.aspx

[29] Solving the E-Waste Problem. Recommendations for Standards Development for the Collection, Storage, Transport and Treatment of E-Waste [Internet]. 2014. Available from: http://www.step-initiative.org/files/_documents/whitepapers/StEP_WP_Standard_20140602.pdf

[30] Silicon Valley Toxics Coalition. Poison PC's and Toxic TV's: California's Biggest Environmental Crisis That You've Never Heard Of [Internet]. 2001. Available: http://svtc.org/wp-content/uploads/ppc-ttv1.pdf

[31] Ceballos D, Gong W, Page E. A pilot assessment of occupational health hazards in the U.S. electronic scrap recycling industry. Journal of Occupational Environmental Hygiene. 2015;**12**(7):482-488. DOI: 10.1080/15459624.2015.1018516

[32] Leung A, Duzgoren-Aydin N, Cheung KC, Wong M. Heavy metals concentrations of surface dust from E-waste recycling and its human health implications in Southeast China. Environmental Science and Technology. 2018;**42**(7):2674-2680. DOI: 10.1021/es071873x

[33] Wong C, Dozgoren-Aydin N, Aydin A, Wong M. Sources and trends of environmental mercury emissions in Asia. Science of the Total Environment. 2006;**368**:649-662. DOI: 10.1016/j.scitotenv.2005.11.024

[34] Shen C, Huang S, Wang Z, Qiao M, Tang X, Yu C, et al. Identification of Ah receptor agonists in soil of e-waste recycling sites from Taizhou area in China. Environmental Science and Technology. 2008;**42**(1):49-55. DOI: 10.1021/es071162z

[35] Ha N, Agusa T, Ramu K, Nguyen T, Murata S, Bulbule K, et al. Contamination by trace elements at e-waste recycling sites in Bangalore, India. Chemosphere. 2009;**76**:9-15. DOI: 10.1016/j.chemosphere.2009.02.056

[36] Wang, F, Kuehr, R, Ahlquist, D, Li, J. E-Waste in China: A Country Report: United Nations University StEP Green Paper Series [Internet]. 2013. Available from: http://isp.unu.edu/news/2013/e-waste_in_china.html

[37] BAN (Basel Action Network) and Silicon Valley Toxics Coalition. Exporting Harm: The High-Tech Trashing of Asia [Internet]. 2002. Available from: http://archive.ban.org/E-waste/technotrashfinalcomp.pdf

[38] Sepúlveda A, Schluep M, Renaud F, Streicher M, Kuehr R, Hagelüken C, et al. A review of the environmental fate and effects of hazardous substances released from electrical and electronic equipments during recycling: Examples from China and India. Environmental Impact Assessment Review. 2010;**30**:28-41. DOI: 10.1016/j.eiar.2009.04.001

[39] Robinson BH. E-waste: An assessment of global production and environmental impacts. Science of the Total Environment. 2009;**408**(2):183-119. DOI: 10.1016/j.scitotenv.2009.09.044

[40] Little PC, Lucier CA. Global electronic waste, third party certification standards, and resisting the undoing of environmental justice politics. Human Organization. 2017;**76**(3):204-214. DOI: 10.17730/0018-7259.76.3.204

[41] Cahill TM, Groskova D, Charles MJ, Sanborn JR, Denson M, Baker L. Atmospheric concentrations of polybrominated diphenyl ethers at near-source sites. Environmental Science and Technology. 2007;**41**:6370-6377. DOI: 10.1021/es070844j

[42] Tsydenova O, Bengtsson M. Chemical hazards associated with treatment of waste electrical and electronic equipment. Waste Management. 2011;**31**(1):45-58. DOI: 10.1016/j.wasman.2010.08.014

[43] Rosenberg C, Hameila M, Tornaeus J, Sakkinen K, Puttonen K, Korpi A, et al. Exposure to flame retardants in electronics recycling sites. The Annals of Occupational Hygiene. 2011;**55**(6):658-665. DOI: 10.1093/annhyg/mer033

[44] Danish Ministry of Environment. The Greening of Electronics [Internet]. 2012. Available from: http://www2.mst.dk/Udgiv/publications/2012/07/978-87-92779-99-1.pdf

[45] Oguchi M, Sakanakura H, Terazono A. Toxic metals in WEEE: Characterization and substance flow analysis in waste treatment processes. Science of the Total Environment. 2012;**463**:1124-1132. DOI: 10.1016/j.scitotenv.2012.07.078

[46] Ceballos D, Dong Z. The formal electronic recycling industry: Challenges and opportunities in occupational and environmental health research. Environment International. 2016;**95**:157-166. DOI: 10.1016/j.envint.2016.07.010

[47] Platform for Accelerating the Circular Economy. A new circular vision for electronics: Time for a global reboot. In: World Economic Forum, Davos 2019; 22-25 January 2019; Switzerland. pp. 1-24

Chapter 5

Wastes from Industrialized Nations: A Socio-economic Inquiry on E-waste Management for the Recycling Sector in Nigeria

Ojiyovwi Johnson Okorhi, Douglason Omotor and Helen Olubunmi Aderemi

Abstract

An "assessment of waste electrical and electronic equipment (WEEE or e-waste) management strategies in Southeastern Nigeria" was conducted towards suggesting appropriate implementable measures. This submission presents a key outcome of a socioeconomic study on factors influencing the paths of e-waste generation and control with a view to suggesting innovative measures and market potentials for firms in the recycling sector. The concept of the study highlighted strategic features in-line with the socioeconomic assessment of e-waste management. Potentials for innovation in e-waste recycling were discussed in-line with elements of sustainability. The research introduced investigative methods by questionnaire administration. Purposive selections of local government areas were made from five mutually exclusive states. Data were analyzed using descriptive statistics. Results revealed the reasons limiting e-waste management trends to include cheap pricing, availability, quality, as well as superiority of obsolete e-devices to newer EEE. Sustainable benchmarks for evaluating and adopting e-waste recycling technologies were recommended.

Keywords: waste electrical and electronic equipment (WEEE or e-waste), recycling, socioeconomic, innovation, southeastern Nigeria, sustainability, technologies

1. Introduction

As one of the main growing waste streams globally [1, 2], a phenomenal increase in the quantities of disposed waste electrical and electronic equipment (WEEE or e-waste) was globally recorded in more than a few parts [3, 4], therefore, seeking for interventions from policy makers and practitioners, as well as the scientific community. The quantity of disposed electrical and electronic equipment (EEE) has been described to increase at a high rate, especially in industrialized countries where markets are flooded with large volumes. Today, the short product lifecycles and rapid innovation in EEE production have resulted in large number of rather new products been thrown away [5, 6]. Estimations places the annual globally volume of generated e-waste to be between 20 and 50 million tonnes [1, 3, 7]. There have been substantial media reports on transboundary movement of WEEE in Nigeria [8]. Records in 2011 shows that Nigeria imported 1.2 million tonnes of new e-devices and generated e-waste of 1.1 million tonnes [9]. With these mounting quantities of WEEE, focus attention is now extended from how WEEE is managed to include reasons for the rising volume and avenues for it to be avoided [10].

Many nations are now faced with the task of handling e-waste that are internally generated and those imported from abroad. Findings revealed that many used electrical and electronics equipment (UEEE) shipments into developing nations are combinations of nearly 25% of disused or end-of-life (E.o.L) e-devices and more than 75% of e-waste [9]. On the contrary, e-waste, though a take on problem, could be an important and alternate source for manufacturing materials whenever it is collected, dispersed and reprocessed properly [10–12]. An entirely new business opportunity is developing with the merchandising, recycling and reprocessing of WEEE [12, 13]. Subjective evidences suggest that there are insufficient actions of management functions for WEEE activities in most emerging nations [7, 9, 14]. Modern trends in recycling of WEEE, still fall short of global practice. Hence, the slow and steady upsurge in the volume of WEEE generated thereby strengthening the concern for waste recovering to protecting valuable materials and safeguarding human health and the environment [1, 13].

Nigeria's approach to WEEE management is seen as considering such emerging waste more on a basis of socioeconomic benefits instead of a long-term human health and environmental effects [7]. Reports by several authors including those by the Öko-Institut and Green Advocacy Ghana in 2010 [15] and Osibanjo and Nnorom [4] revealed that this is driven by an approach to catch-up with the "digital divide" through imports of low-priced near E.o.L EEE from industrialized countries. Many E.o.L e-devices are reasonably stockpiled instead of direct disposal with everyday household refuse [16]. Policy regulators and monitors at the local government areas (LGAs), whose mandates covers solid waste management [17], have unsuccessful establish workable management policy for e-waste management [10]. In several industrialized countries with workable policy frameworks for e-waste, there have raised new businesses revolving around tradeoff, reprocessing and repairs of E.o.L EEE [18]. Primarily, this has been linked to the huge volume of precious metals found inside e-waste. The ratio of prized metals to waste in various E.o.L EEE (especially iron, aluminum, copper, gold etc.) is found to supersede its associated pollutants, therefore encouraging recycling in the e-waste sector [19]. Therefore, the study assessed the socioeconomic factors swaying the paths of e-waste generation and control in Southeastern Nigeria with a view to suggesting innovative measures and market potentials for firms in the recycling sector.

2. WEEE streams: The trail to developing nations

The main sources for WEEE inflow into Nigeria is mapped out to include the container market and RoRo market [18]. It is estimated daily that 500 containers of used laptops, computers, televisions and other e-devices are imported into Nigeria Ports [9]. The 2011 Basel Report noted that e-waste comes to Africa predominately from Europe (majorly through the ports of Felixstowe, Amsterdam and Antwerp). The Nigerian counterpart, Belgian Customs estimates that nearly 90% of these prohibited shipments influx Nigeria environment from co-loaded automobiles with E.o.L EEE [7, 18]. On inspection, many of the exports have problematic contents or are in fact mislabeled for ease of shipment of what are in fact illegal goods. In 2008, the inspection of containers by the Nigeria Customs led to the discovery of 127 e-waste containers, from which 47 of them considered hazardous were reversed and shipped to the origin sources abroad. Up to 2011, the National Environmental Standards and Regulations Enforcement Agency (NESREA) had impounded five vessels carrying WEEE destined for Nigeria [9].

2.1 Bearings of the WEEE value-chain in Nigeria

e-Waste management in Nigeria is now been tackled not only by relying on prohibiting illegal imports, but by embracing other management strategies through the processes of generation, collection, handling, recovery, recycling and towards final disposal. In this regard, e-waste is considered with the idea of evaluating consumer's behavioral and its socioeconomic implications [15, 18]. Following the Basel Report on e-waste for 2011, a well-coordinated cluster of e-waste recyclers in some regions across West Africa focused their collection activities mainly on UEEE (or end-of-life EEE) and discarded e-waste. These traders source the items from locally generated and foreign imports which is based on categories of items been savaged [7]. The report showed that those in the recycling sector are engaged in recovering e-waste from waste streams, worked on these items and recovered several types of components and materials. Such recovered components from disassembled devices sometimes sever as sources for repair spare-parts. In another report, Lagos, Nigeria has two main recycling clusters located at Alaba International Market and Ikeja Computer Village employing nearly 15,000 technicians and traders with more than 5000 registered enterprises [10]. These two locations were characterized with high patronage by Nigerians, as well as West and Central African nationals in the sales and professional repairs of refurbished EEE.

Furthermore, it is on record that the collection, handling and refurbishing of e-waste in Nigeria take place mainly in the informal sector of recycling by inexperience, low-class, illiterate and undocumented-business individuals. Some of these scavengers, with no prior training and little investment, move around neighborhoods and waste dumps with their handcarts to collect (or in some cases buy) disused e-devices and related metal scraps that contain valuable like aluminum, copper, brass, iron, etc. [7]. These items recovered are then sold directly to cottage recycling businesses (engaged in dismantling to recover valuable components) or to secondary traders that organize large-scale sales to local and foreign recycling firms [10]. The remnant from the dismantled items is often subjected to indiscriminate disposals - including burning (especially plastics coated materials) [8, 16]. Besides, these scavengers are guaranteed of steady access to daily pay, as the proceeds from each day's scouting immediately materialize on sales of the recovered components.

2.2 Pathways for e-waste generation and recycling of the households and traders

The transboundary movement of UEEE/WEEE in industrialized nations varies from one country to another. In certain instances, private households organize their e-waste disposal by requesting either government service or private service, usually for a price [1]. More often, the scheduled bulky waste pick-up service is managed by private collectors who are often concerned with the afterward segregation of the collected wastes towards recovery and recycling. Whereas, the measures used during "recycling" in Nigeria are comparably considered crude and unstandardized. Recovered components are sometimes sold for export to other places in Africa and Asia [7].

The transboundary movements of e-waste in West Africa countries is found to be driven by a craving for UEEE/WEEE owing to its cheap pricing, quality and durability [7, 8]. The brokers and traders of WEEE have been identified as some of the key players in this trade. This sector ranges from household-arrangement to a bulky and well-arranged distributing syndicates. They are well organized and linked from their point of shipment (Europe) to destination (Africa). Another influence in the collection and handling of UEEE/WEEE is the recurrent visits of WEEE traders to designated formal collection centers to request certain useful items for free with a view to process them for export outside Europe [1, 20]. In this vein, some amount of e-waste somehow finds its way into informal arrangements. Consequently items originally designated for recycling plants are diverted from the formal value-chain into the informal sector. The Basel Report of 2011 stressed that this trail of UEEE from the formal recyclers to informal recyclers is the deviation orchestrated by some registered middlemen to illicit traders (or informal recyclers). These brokers act as logistic firms or sub-contractors in scheduling pick-up services for WEEE, and in many instances consent certain items requested by WEEE traders from the waste streams [7]. In turn losing track of what becomes of such items at the end.

2.3 The nexus of e-waste and the SDGs

As e-waste recycling scheme is gaining more attention, there remains slow competitiveness for the adaptation of innovative technologies in the preprocessing of WEEE. Hence, there is a strong need for the adoption of frontier technologies in recycling. Consequently, the problems of WEEE could be linked to the sustainable development goals (SDGs) in areas of building strong businesses, promoting inclusive and sustainable industrialization and fostering innovation (goal-9); justifiable economic growth, complete and productive employment with decent working environment for everyone (goal-8); as well as maintaining sustainable resources usage and production patterns (goal-12) [21]. Therefore, sustainable management of e-waste in Nigeria and its possible recycling is of high relevance to the SDGs—the planet goals—especially to the prosperity goals, and particularly to goal-12.

2.4 Theories and concepts for the socioeconomic evaluation of e-waste management

The old perception of waste disposal—"dilute and disperse" is no longer tenable, rather a novel model of "concentrate and contain" has paved way to an idea referred to as the "Integrated Waste Management Scheme" [16, 22]. Generated waste is now deliberated as wealth out of place. Numerous waste items can now be collected, refurbished, and reused in the industries, agricultural, construction and building sectors etc. thereby safeguarding natural resources and energy in production of new items. Such measures also minimize environmental effects and relative health issues that could arise from the continuous exploitation of natural resources [10, 23]. This study is driven by the Pongrácz "theory of Waste Management" which is grounded on an agreed expectation that waste management can prevent waste to safeguard man and his environment. It assumes that the practice of waste management would avoid resources losses by turning waste to resources and conserving natural resources. Hence, the theory suggests that "we shall prevent waste from being produced by producing useful products (non-wastes) primarily" [11].

Hence, a conceptual framework (**Figure 1**) was developed to address the socioeconomic factors for the sustainable management of WEEE in the Recycling Sector. The fundamental aspects in this e-waste framework include the "Political, Institutional, Social, Financial, Economical and Technical". There are four contextual concerns raised in WEEE management which are namely: "Environmental, Sociocultural, Political and Economic" [10, 16, 24].

Individuals' behavior and approach to managing their generated wastes differs owing to their social and cultural traits. For example, people living in a fast

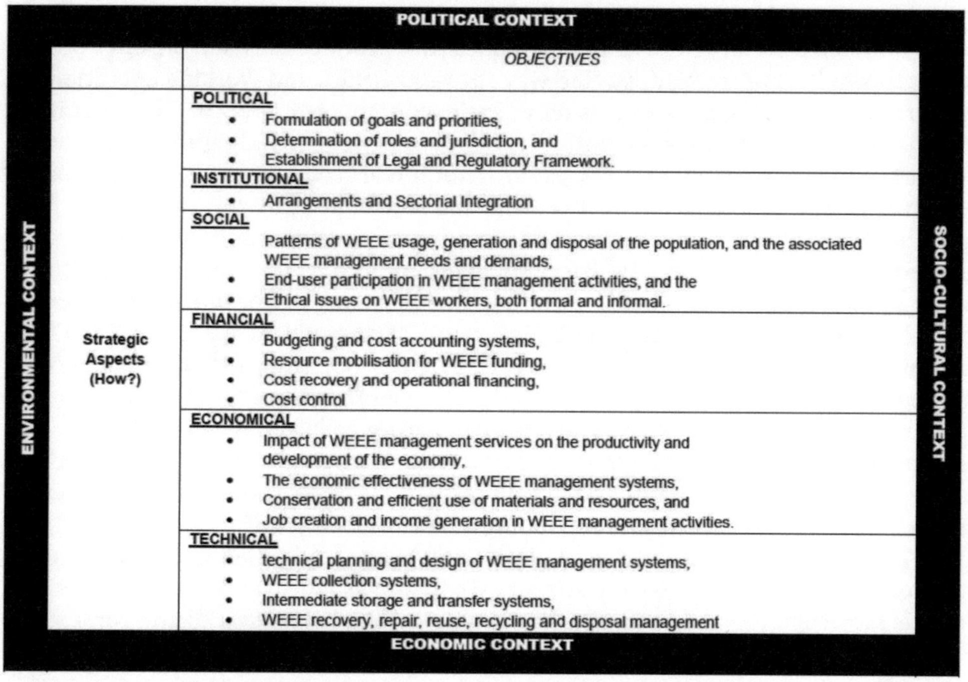

POLITICAL CONTEXT

ENVIRONMENTAL CONTEXT

SOCIO-CULTURAL CONTEXT

	OBJECTIVES
Strategic Aspects (How?)	<u>POLITICAL</u> • Formulation of goals and priorities, • Determination of roles and jurisdiction, and • Establishment of Legal and Regulatory Framework.
	<u>INSTITUTIONAL</u> • Arrangements and Sectorial Integration
	<u>SOCIAL</u> • Patterns of WEEE usage, generation and disposal of the population, and the associated WEEE management needs and demands, • End-user participation in WEEE management activities, and the • Ethical issues on WEEE workers, both formal and informal.
	<u>FINANCIAL</u> • Budgeting and cost accounting systems, • Resource mobilisation for WEEE funding, • Cost recovery and operational financing, • Cost control
	<u>ECONOMICAL</u> • Impact of WEEE management services on the productivity and development of the economy, • The economic effectiveness of WEEE management systems, • Conservation and efficient use of materials and resources, and • Job creation and income generation in WEEE management activities.
	<u>TECHNICAL</u> • technical planning and design of WEEE management systems, • WEEE collection systems, • Intermediate storage and transfer systems, • WEEE recovery, repair, reuse, recycling and disposal management

ECONOMIC CONTEXT

Figure 1.
Conceptual framework for WEEE management strategies (Adapted from Okorhi [9], assessment of WEEE management strategies in South Eastern Nigeria).

developing low-income community have been found to constitute a majority of diverse indigenous group with social difference gap. With such gap of the populace, organizing a thorough e-waste management at such places would be challenging. Secondly, the concern of lobbyists, interest groups and political parties would definitely affect to a large extent the kind of management strategies that is finally put in place for managing e-waste in a community [24]. Therefore, there is a need to incorporate in every stage of the policy making process individuals' views and participation. Lastly, the purpose of e-waste management, its technical and organizational scheme would depend in general on both the economic context of the inhabitants and the economy of the town. For example, in some fast developing towns like Enugu, Onitsha and Aba in Southeastern Nigeria, there are renowned specialized markets boosting informal trade in Fast Moving Consumer Goods with high trade volumes [25], and its highly characterized waste management problems [26]. Consequently, the level of economic development is a vital factor in the amount and composition of e-waste generated in that place [27]. Therefore, to accomplish the objective of this paper, the authors focused on assessing the strategic aspects revolving the political structure, social context, individual economy and technical inputs.

2.5 Market potential of innovative e-waste recycling at firm level

The industrialization of Africa could be achieved through sustainable innovation and awareness creation of its innovation potentials. According to a report published by Schluep et al. in 2009 [28], sustainable innovation refers to the shift of sustainable technologies, products and services to the marketplace, requiring a market creation concept and a shared global agenda. Whereas, environmental management and sustainability focuses on finding solutions to global pressing environmental problems. It is said that the best available environmentally sound management (ESM) systems are programs and techniques that produces sustainable environment through its protection, paving way for safer health and working conditions, generating employment as well as other socioeconomic benefits [10]. In pursuit of these, there arises the deployment of frontier strategies (including the 5Rs) in e-waste management. However, the activities of metal recyclers in Nigeria are secondarily connected with the e-waste recycling sector, because the business outputs are a measure of functional items and valuable components rather than just raw materials [7]. Though, the 2011 Basel Report found the sector producing significant amount of e-waste. This is because the e-waste recycling sector in Nigeria is dominated by firms (or individuals) with "informal" arrangements which collect WEEE at random, manually dismantling (or sorting), preprocessing, selling valuable components and, disposal of the leftovers [10]. On the other hand, prized metals present in printed wiring boards (PWBs) are hardly collected for export to recycling facilities, and when that happens, the selling price is often below world market prices and discouraging to WEEE traders [7]. Also, some devices extracted from WEEE are used as spare parts in the repairs of faulty EEE.

Obviously, the ease to getting vital production materials used in the manufacture of new EEE is progressively attracting concern as global reserves of raw materials is fast declining and becoming more expensive [1]. The overall aim for "formal" e-waste recycling is to avert hazardous materials from WEEE in an ESM manner; recover prized items as much as possible; build an eco-friendly and sustainable SMEs and; consider the socio-economic implications [24]. Consequently, the recycling of e-waste is a key strategy for reducing "stockpiled" waste streams, minimizing the consumption of natural resources as well as improving energy usage. In this light, the paper briefly discuss the sustainability benchmarks for evaluating and adopting technologies for e-waste recycling; some innovative WEEE recycling technologies that could be adopted by recycling firms; as well as the market potential for e-waste recycling in many developing nations.

For a better consideration of the procedure for selecting innovative e-waste recycling technologies in developing countries, Schluep et al. [28] suggested, among others, the importance of sustainability benchmarks. **Table 1** shows the sustainability benchmarks for evaluating and adopting technologies for WEEE recycling in developing nations, including Nigeria. The benchmarks to compare the innovation of technologies were then grouped with elements of sustainability. Whereas, **Table 2** shows some innovative e-waste recycling technologies that could be adopted by e-waste recycling firms in Nigeria.

To sum up, the market potential for e-waste recycling are enormous as the annual growth rate of WEEE in Nigeria is put at 10% in the volume of waste generated [13]. It has been identified that a mid-term medium potential for integrated e-waste smelting already exist in some countries of Asia, Africa, South and North America [24]. Hence, from job creation, entrepreneurship and sustainability viewpoints, the "informal" practices of collection and manually dismantling of e-waste may not really require a transformation to a "formal" arrangement using high-tech equipment for the processing of WEEE [29]. The innovative technologies been continuously adapted by the larger informal sector in Nigeria is gaining ground [7]. Opportunities in recycling of e-waste arise in the improvement of the processing of cable-coated from poly-vinyl-chloride and insulators, and poly-brominated biphenyls coated plastics. Also is the collection of large quantities of PWBs for export and fair pricing. By using the voluntary carbon standard (VCS) or carbon action reserve (CAR) schemes, there is now the potential of recovering chlorofluorocarbon from cooling units and insulation foam which in turn brings both environmental and economic gains [7]. It was also suggested that the improved

Attributes	Indicators involved
Economic attributes	
Low net costs	Costs for transport, processing and labour vs. revenues
Low capital costs	Investment costs for additional plants and technologies used in a scenario
Increased potential for local economic growth	Additional industries and services involved by implementing a scenario
Environmental attributes	
Low use of electricity	Savings of electricity but also energy in general by implementing a scenario
Low fuel use for transport	Fuel used by shipping and road transport
Low use of freshwater	Freshwater consumption of a recycling scenario
Little (toxic) emissions	Caused vs. prevented emissions according to the savings of raw materials calculated with eco-indicator '99 (or other appropriate tools)
High metal recovery rates	Range and yields of metals contained in the waste, which can be recovered and used as secondary raw material. In case of technical conflicts prioritization by economic and environmental value ("footprint") of the recovered substances.
Social attributes	
Creation of jobs for the previously unemployed	Working hours for low-skilled and semi-skilled workers generated
Creation of highly skilled jobs	Working hours for highly skilled workers generated
Creation of jobs outside the target country	Working hours generated outside the target country
Low health and safety impacts	Impacts of a scenario on health and safety of the employees engaged in a scenario

Table 1.
Sustainability benchmarks for evaluating and adopting technologies for e-waste recycling in developing countries (Adapted from Schluep et al. [28]).

	Waste streams	Economic attributes	Environmental attributes	Social attributes
Manual dismantling/ sorting of fractions	All	Low capital cost, sorting of valuable fractions/ components	Efficient sorting of fractions	Labour intensive, Job creation
De-gassing CFC, HCFC	C&F	Mandatory requirement having low cost	Fundamental step to ensure control over hazardous substances having huge GWP potential	
Semi-automatic CRT cut and cleaning	CRT	Low capital and net cost	Low energy consumption	Labour intensive
Integrated smelter for non-ferrous (pyrometallurgical methods)	Non-ferrous (including printed wiring boards) like Cu, Pb, Zn, Sn or mix	Capital cost high Low net (unit) costs due to economies of scale Local growth potential high	No toxic emissions Low water use Transport: internationally Little waste products Recovery rates >> 90%	Automated process control so less jobs created Highly skilled workforce EHS*
Aluminium remelter/refiner	Aluminium	Captial cost medium – high Net cost low Economies of scale	No toxic emissions Salt slag has to be treated or disposed Env. sound Transport within region or country Water use: low - medium	Job creation: yes Mix of low skilled and high skilled jobs EHS low risks

Table 2.
Innovative e-waste recycling technologies for recycling firms (Adapted from Schluep et al. [28]).

utilization of polystyrene from e-waste recycling would guarantee a higher monetary value from the pricing of carbon (IV) oxide.

3. Materials and methods

The methodical conception for this article is based on both reviews from available literature on sustainability, innovations and management strategies for e-waste, and results from e-waste survey carried out in Southeastern Nigeria. The survey was carried out in mutually exclusive strata of States (Enugu, Anambra, Ebonyi, Imo and Abia) with 95 local government areas (LGAs). A local government area (LGA) was purposefully chosen from every senatorial district in each State, and the fourth LGA was taken as the State's capital. This selection was predicated on the high volume of e-waste generated and handled in these LGAs. Altogether, 20 LGAs were picked for the survey, with 4 LGAs selected from each State. Questionnaires were administered in each of these LGAs to end-users, traders/recyclers, and policy makers/monitors assessing the "socioeconomic of WEEE" of the study area. A population of 280 respondents was surveyed. From the table for determining sample size [30], a population of 280 respondents gives 162 sample size representatives of the respondents. Using qualitative and quantitative methods, the study measured e-waste needs and demands; generation, collection and final disposal; recycling measures; associated jobs; incomes to traders and technicians; as well as technology frontiers. Both **Table 3** and **Figure 2** shows methodological approach employed. The analysis that followed established the extent of reliability and a 95% confidence level placed on the information elicited.

Sketchy findings suggest that a sustainable WEEE management scheme requires sufficient and continuous financing, frontier technologies, an equipped working environment and the right institutional motivations for key players [10]. The end-users of WEEE are mainly responsible for the patronage/usage of these UEEE. WEEE traders (or recycling firms) serve to collect and distribute these items, while the monitoring agencies ensure execution of policies, taking feedbacks and acting as check on other stakeholders. Hence, we administered three distinct questionnaires to these three players. **Table 4** shows an outline of the main areas of interest captured in the respective questionnaire.

Three distinct questionnaires were administered to stakeholders. These are (1) the policy regulators and managers vis-a-vis "NESREA, SON, State Environment Protection Agencies, Environmental Health offices & Nigeria Customs Service"; (2) e-waste traders/recyclers—"Dealers, Marketer, Retailers, Technicians and Refurbishers/Recyclers of WEEE"; and (3) e-waste consumers/end-users—"Households, Government Institutions, Industries, Private Offices, Communication/Entertainment Businesses, Educational and Health-Care Centers".

The responses are presented in tables and figures in the sections that follow. The tables depict a collection of these stakeholders, managerial framework and end-users' participation. It then measured waste disposal pattern by the consumers, as

Stakeholders	Number administered	Number retrieved	% of Number retrieved	Number of valid retrieved questionnaire	% of valid retrieved questionnaire
Monitoring/ control agencies	40	40	100	35	12.50%
Distributors/ recyclers	40	29	72.5	29	22.86%
Consumers/ end-users	200	137	68.5	137	48.93%
Total	280	206	73.6%	201	71.79%

Source: Field Survey, 2015.

Table 3.
Schedule of questionnaire administered.

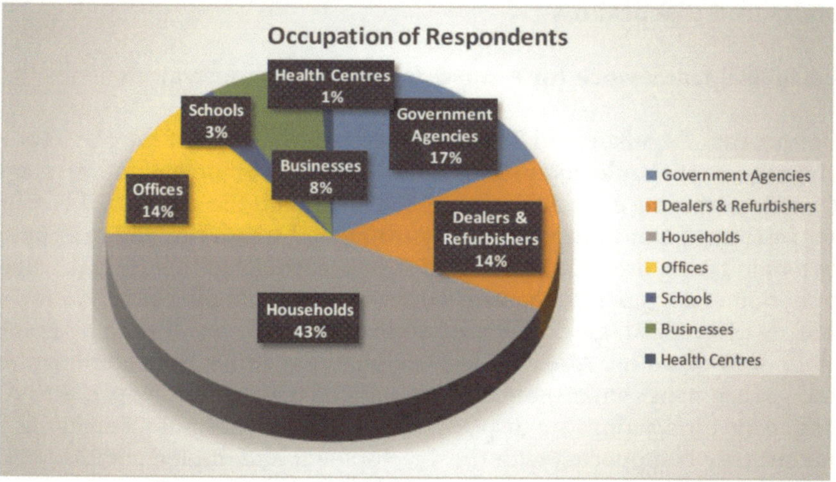

Figure 2.
Representation of the retrieved administered questionnaire.

S/No	Outline of key areas of interest of the questionnaire for "Socio-economic assessment of e-waste"	Stakeholder(s)	Questionnaire type administered	No. of respondents
1	Administrative framework for WEEE	Monitoring/control agencies	A	35
2	End-user participation in WEEE management activities	Monitoring/control agencies	A	35
3	e-Waste disposal practices and patterns by the consumers/end-users	End-users/consumers	B	137
4	Factors considered in adopting strategies for final disposal of WEEE by recyclers and dealers	Distributors/recyclers	C	29
5	WEEE collection, handling and disposal methods by entrepreneurs and recyclers in South Eastern Nigeria	Distributors/recyclers	C	29
6	Socio-economic drivers on trends in generation, collection and disposal of WEEE	Monitoring/control agencies; end-users/ consumers; distributors/ recyclers	A, B and C	201
7	Factors influencing technical planning and design for WEEE management systems	Monitoring/control agencies	A	35

Table 4.
Key sections of the questionnaire.

well as factors leading to choice of strategies adopted for the final disposal of e-waste recyclers and traders. It showed some of the strategies used by these entrepreneurs in the gathering, management and final disposal of WEEE. Lastly, it reflected on the socioeconomic drivers of e-waste, and the issues militating on sustainable framework for e-waste management systems.

4. Results and discussions

4.1 Managerial framework for e-waste by policy administrators

Government's Regulation S.I.28 of 2009 stresses that part of the plans for e-waste management should comprise endorsing current guidelines and strategies for "solid waste (including e-waste) management" through the conduct of baseline surveys, instituting public health and environmental standards, and making sure there is a monitoring program that include early warning system [6, 10]. Therefore, the promotion of a supportive management strategies and plan of action for WEEE was confirmed by 25(71.43%) policy administrators to be an all-encompassing aspect of policy planning. Also, **Table 5** revealed that 19(54.29%) administrators affirmed government's endorsement of regulations which enforces protection to the environs against indiscriminate disposal of e-waste. Furthermore, results showed that this strategy is supported with the development and implementation of strategic work plans for this special waste in assisting stakeholders—23(62.71%) of these respondents. Nonetheless, 8(22.86%) administrators affirmed that policy regulators (tiers of government) put in place dedicated and competent bodies to implement strategies for e-waste management, while nearly half of these officials—15(42.86%) argued that such a specialized section or unit for an exclusive management of e-waste do not exist in their establishments. Field observations showed that in few places where such relative departments exited, it was rooted under units such as "special waste unit" or "harmful waste division", and it barely gets adequate appropriations to combat these special wastes. Lastly, results revealed that the management strategies for operation were considered unsuitable by a total of 25 (71.43%) policy regulators (from combining 8(22.86%), 13(37.14%) and 4(11.45%) respondents).

4.2 Regulators opinion of end-users' participation in e-waste management activities

Public education and participation are necessary to support the plan of action for e-waste management. This is in order to achieve an efficient implementation process of management strategies. As depicted in **Table 6**, together 18(51.43%) policy regulators affirmed government engagement in the sensitization of interested parties. While more than half of them—19(54.29%) agreed that the populace is amply involved with the implementation process of control strategies. However, 24(68.57%) respondents admitted that end-users simply comply with the implemented strategies. Furthermore, these administrators also suggested that the common means for e-waste sensitization were executed with Radio jingles/programs—17(48.57%), Television announcement/documentaries 5(14.29%), Posters 5(14.29%), Handbills/flyers 5(14.29%), as well as (mobile advertisements, campaigns, road-shows, etc.) 3(08.57%) respondents.

4.3 Disposal practices and patterns of e-waste by the end-users

Together **Table 7** and **Figure 3** underlined management strategies adopted and practiced for e-waste by the customers. This assessment discovered that the most common strategy adopted in the final disposal of WEEE by many homes and businesses is the direct disposal of e-waste along with other regular solid wastes—96 (70.07%) respondents. Additional measures embraced by the households included the reselling of disused EEE—32(23.36%), and stockpiling—21(15.33%). In few

	Q/N	Policy instrument: Does your agency:	To very great extent		To great extent		To small extent		To very small extent		Not at all	
			N	%	N	%	N	%	N	%	N	%
Political framework for WEEE	1	Provide e-waste management tenets in written codes	4	11.45	4	11.45	5	14.29	5	14.29	17	45.57
	2	Prepare and develop working and management plans to stakeholders	6	17.14	17	45.57	10	28.57	1	02.86	1	02.86
	3	Have adequate periodic documentation on e-waste quantity and budgeting to support management process	1	02.86	3	08.57	3	08.57	10	28.57	18	51.43
	4	Establish a competent body to implement e-waste management strategies	4	11.45	4	11.45	8	22.86	4	11.45	15	42.86
	5	Monitor the sources of e-waste into South Eastern Nigeria	2	05.71	8	22.86	5	14.29	6	17.14	14	40.00
	6	Monitor and protect the environment against illegal e-waste dumping	9	25.71	11	31.43	3	08.57	4	11.45	8	22.86
	7	Promote strategies/ policies/legislations/ acts/regulations for WEEE management	6	17.14	19	54.29	8	22.86	0	00.00	2	05.71
	8	Promulgate edicts to enforce protection policies against illegal disposal of WEEE	7	20.00	12	34.29	2	05.71	7	20.00	7	20.00
	9	Enact appropriate legislation on grading rules, waste minimization and so on to back e-waste management strategies	6	17.14	13	37.14	8	22.86	2	05.71	6	17.14
	10	Are the strategies for implementation appropriate?	2	05.71	8	22.86	8	22.86	13	37.14	4	11.45

Total number of respondents = 35.
Source: Field Survey, 2015.

Table 5.
Administrative framework for WEEE.

cases, end-users were found to abandon their defective e-devices with technicians/recyclers who at times refurbish or recovers valuable components—21(15.33%) end-users. In similarly manner, some consumers take apart components of simple devices and reclaim functional parts—25(18.25%). Also, end-users admitted donating certain disused devices to individuals, friends, religion centers, schools, non-governmental organizations (NGOs), etc.—19(13.89%). Besides, it was shown that

	Q/N	Policy instrument: Does your agency:	To very great extent		To great extent		To small extent		To very small extent		Not at all	
			N	%	N	%	N	%	N	%	N	%
Public education and participation	1	Educate the public on e-waste management scheme	7	20.00	11	31.43	2	05.71	2	05.71	13	37.14
	2	Are all sectors of the populace adequately carried along during implementation of strategies?	2	05.71	9	25.71	14	40.00	5	14.29	5	14.29
	3	Does all sectors always comply with the strategies employed?	0	00.00	2	05.71	9	25.71	15	42.86	9	25.71

Total number (N) of respondents = 35.
Source: Field Survey, 2015.

Table 6.
End-user participation in WEEE management activities.

whatsoever strategy choice(s) chosen by the consumer, the state of the E.o.L EEE or e-waste was definitely taken into account before disposal. 100(72.99%) end-users said that their e-devices which were damaged beyond repairs would certainly be thrown away. However, 4(02.92%) consumers agreed that they would rather throw away any disused EEE which could likely be repaired. An additional 33(24.09%) respondents proposed that E.o.L EEE or disused (obsolete) EEE would also be thrown into the waste stream (**Figure 3**). In addition to the aforementioned decisions, 91(66.42%) consumers established that they hardly apply any particular stratification measure for generated e-waste before the final disposal into waste streams. Specifically, 85(62.04%) end-users confirmed that their disused batteries are disposed along with other household waste.

4.4 Factors influencing the adoption of strategies for the disposal of e-waste by recycling firms

Starting with **Table 8**, several factors were admitted by the stakeholders as reasons for the choice of final disposal of generated e-waste. The survey considered some of these drives to include: obsolescence devices; damaged beyond parts; high cost of maintenance/replacement of components; unavailable spare-parts; as well as unwarranted e-devices. Additional reasons considered by the respondents included business growth, innovation within the firm, slow processing speed of e-devices, inadequate storage capacity of EEE, faults from power-surge, and fault resulting from lightning. Field survey results [10] showed that many recyclers/technicians throw away disused e-devices owing to outdated functionality—12(41.38%), and when these items are broken beyond repair—15(51.72%). One more noteworthy cause for this latter practice is the absence of replacement spare-parts—9(31.04%) respondents. On the other hand, e-waste traders were unlikely to dispose of faulty e-devices because of non-warranty (divestment)—9(31.04%); business expansion—9(31.04%); power-surge faults—10(34.48%); as well as damages occasioned by lightning—12(41.38%). These second factors are the obvious reasons for e-waste stockpiling in my places and locations surveyed.

Q/N	(Section D—Consumers/end-users questionnaire) Policy instrument		N	%
	Question	**Option**		
20	How do you discard your waste electronics devices?	Keep in store room	21	15.33
		Resell the devices	32	23.36
		Disposed with general waste	96	70.07
		Give them to a recycler	21	15.33
		Donate to family, friends, school, NGO, etc.	19	13.89
		Return to the store where it was bought for a reduction on the price of a new device	11	08.03
		Return to the seller on a buy-back arrangement	2	01.46
		Disassemble to reuse some parts	25	18.25
		Put it on the street	2	01.46
		Give it to hawkers	1	00.73
21	At what state do you do this?	Broken—Not repairable	100	72.99
		Broken—repairable	4	02.92
		Old or out dated (Obsolete)	33	24.09
27	Do you apply any specific classification/stratification for e-waste before disposal?	Yes	15	10.93
		No	91	66.42
		Not Sure	31	22.63
28	How do you dispose used batteries?	Disposed along with other waste	85	62.04
		Stratified and disposed alone	18	13.14
		Disposed along with other classified hazardous waste	21	15.33

Total number (N) of respondents = 137.
Source: Field Survey, 2015.

Table 7.
e-Waste disposal practices and patterns by the consumers/end-users.

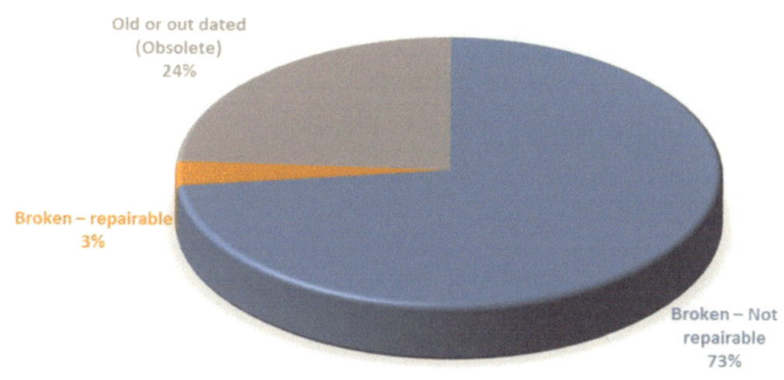

Figure 3.
Disposal measures adopted by the end-users for WEEE generated.

S/N	Factors	Ranking									
		Insignificant ↔ Most_Significant									
		1		2		3		4		5	
		N	%	N	%	N	%	N	%	N	%
1	Functional obsolescence	4	13.79	2	06.70	5	17.24	6	20.67	12	41.38
2	Damage beyond repair	2	06.70	2	06.70	4	13.79	6	20.67	15	51.72
3	Cost of maintenance	4	13.79	8	27.59	7	24.14	7	24.14	2	06.70
4	Repair components not available	3	10.35	0	00.00	6	20.67	4	13.79	9	31.04
5	Divestment	9	31.04	9	31.04	4	13.79	3	10.35	4	13.79
6	Expansion of business	9	31.04	9	31.04	3	10.35	5	17.24	3	10.35
7	Business innovation	6	20.67	8	27.59	7	24.14	5	17.24	3	10.35
8	Processing speed inadequate	8	27.59	3	10.35	11	37.93	5	17.24	2	06.70
9	Storage capacity inadequate	7	24.14	8	27.59	6	20.67	6	20.67	2	06.70
10	Power surge	8	27.59	10	34.48	6	20.67	3	10.35	2	06.70
11	Lightning	12	41.38	4	13.79	6	20.67	2	06.70	5	17.24

Total number (N) of respondents = 29.
Source: Field Survey, 2015.

Table 8.
Factors considered in adopting strategies for final disposal of WEEE by recyclers and dealers.

4.5 e-Waste management measures adopted by recycling firms

Table 9 suggests that the stakeholders involved in WEEE refurbishing and recycling applies one or more of the seven standard strategies in the management of generated e-waste. Many entrepreneurs and recyclers of WEEE in Southeastern Nigeria manage their E.o.L EEE and e-waste by adopting strategies like Reuse of e-waste—18(45%); Repair of disused devices—16(40%); and Incineration (burning)—16(40%). In other occasions, technicians searched for and recycle peculiar components from disused e-device that are valuable and could serves as repair spare-parts for other faulty appliances. In such cases, e-waste is dismantled to retrieve these valuable components and reuse directly during repairs or indirectly in developing of new items. A computer technician confirmed the use of Light Emitting Diode

Q/N	(Section D) Policy instrument	N	%
1	Recycling of e-waste	13	32.50
	Reuse of e-waste	18	45.00
	Recovery of e-devices	10	25.00
	Source reduction of generated e-waste	5	12.50
	Repair of E.o.L electrical/electronic equipment	16	40.00
	Landfill of waste	11	27.50
	Incineration of waste	16	40.00

Total number (N) of respondents = 29.
Source: Field Survey, 2015.

Table 9.
WEEE collection, handling and disposal methods by entrepreneurs and recyclers in South Eastern Nigeria.

salvaged from disused laptops in the development of electricity detector used in homes. Likewise, some mobile phone businesses in major commercial towns in Southeastern Nigeria were engaged to recalling E.o.L mobile phones on behalf of the parent manufacturers. For instance, two sales outlets of a particular firm in Enugu metropolis accepted from their customers E.o.L mobile phones as trade-in for a new ones with an average of 70% price (of the new product) being committed by the customer. This is apparent under agreed conditions dictated by the fronting firm to the end-users. In contrast, it is important to mention that the assertion of using incinerators as a strategy in managing e-waste is far from reality in the study area. Observations from the study area revealed that stakeholders rather practiced surface burning of WEEE and this takes place in a number of locations (mostly in low-lying lands). This is clearly misjudged as incineration of e-waste. Despite the fact that both processes lead to combustion of the waste materials, surface burning occurs in lower temperatures of between 20°C and 300°C, and incineration involved higher temperatures ranges of up to 1000°C in an environmentally confined engineered plant that traps ashes and non-combustibles remnants [22]. Not a single stakeholders surveyed possesses or operates a confined incinerator for the aim of e-waste management. Also, 11(27.50%) respondents admitted that generated e-waste was management by landfilling. Yet again, observations on the field suggested otherwise. Similarly, landfills are well-engineered facilities designed, operated, carefully monitored, and located off town. They are closely cared for even after years of closure. It could be cleaned up when need be and pay for to insure adequate compliance with standard environmental laws. From global perspective, several landfills maintenances are intermittently managed by government's prescribed environmental authorities. Most of the surveyed policy regulators could not affirmed to have a well-engineered system of landfill and incinerator in place in Southeastern Nigeria. Also, where claims of landfilling practices took place in the surveyed area, it was another misrepresented for a long term low-land repossession by using collected wastes as a feedstock.

4.6 The socioeconomic drivers on trends in the management of e-waste

Four factors were recognized and reflected as likely economic drivers which determined the disposal pattern of obsolete EEE (or e-waste) in Southeastern Nigeria. These included cheaper e-devices, access to EEE, crave over inferior devices, and the quest for superior EEE. **Table 10** showed officials of the regulatory agencies in the surveyed area strongly affirming some of these key economic drives as access to e-waste—25(71.43%), as well as low-priced WEEE—16(45.71%). From this, a line can be drawn from several literatures which have shown clear suggestions buttressing the claims that Nigeria was undergoing rapid ICT revolution in recent years [13]. As a result to connect with the "digital divide", attempts were made by individuals and e-waste traders to import cheap and (sometimes) durable E.o.L e-devices (or e-waste) from developed countries into Nigeria. Also, **Table 10** showed that the upsurge in the demand by end-users and e-waste traders for UEEE (or e-waste) could be linked to its cheap pricing—141(79.66%); device durability—96 (54.24%); economic class of consumers—77(43.50%); EEE accessibility—82 (46.33%); as well as the quality of WEEE and its superiority to (some brand) new products. While the noting the factors influence the final disposal of e-waste, these respondents associated these to high cost of disposal—43(24.29%); inadequate storage space—38(21.47%); associated disposal fees—46(25.99%); quick obsolesce of UEEE—42(23.73%); and the inaccessibility to formal recycling plants/facilities, as well as enormous cost in setting up a formal recycling facility for e-waste disposal. Owing to such associated cost, there exists only one eco-friendly electronic waste recycling company—E-Terra in Nigeria.

S/N	Question	Responses				
	(Monitoring Agencies = 35)	**Cheap EEE**	**Availability of EEE**	**Inferior EEE**	**Superior EEE**	**Others**
1	What are the economic drives that help to determine the disposal of used electrical/ electronic device?	16 (45.71%)	25(71.43%)	13(37.14%)	5(14.39%)	
	(End-users + Dealers = 177)	**Cost**	**Durability**	**Income**	**Accessibility**	**Others**
2	What give rises to the attractiveness of used (Tokunbo) electrical electronic equipment in South Eastern Nigeria?	141 (79.66%)	96(54.24%)	77(43.50%)	82(46.33%)	Quality(2), superiority, cheap
	(End-users + Dealers = 177)	**Cost of disposal**	**Lack of storage space**	**Money exchanged for WEEE**	**Obsolesce**	**Others**
3	What are the possible economic drivers for final disposal of WEEE	43 (24.29%)	38(21.47%)	46(25.99%)	42(23.73%)	Availability of recycling facilities, cost of recycling (2)

Total number (N) of respondents = 201.
Source: Field Survey, 2015.

Table 10.
Socio-economic drivers on trends in generation, collection and disposal of WEEE.

4.7 Factors swaying the planning and design for sustainable e-waste management systems

The laws and guidelines that support e-waste management schemes in South-eastern Nigeria were identified to be anchored on four strategic aspects and therefore considered for this study. These included establishment of state-of-the-art technologies and essential working equipment, capable and sufficient manpower, funding of WEEE schemes, as well as impediment in implementation of e-waste regulations. From **Table 11**, 18(51.43%) of the monitory and regulatory agencies

S/N	Question:	Are there any particular difficulties in the implementation process of e-waste management strategies?			
		Strongly Agreed	**Agreed**	**Disagreed**	**Strongly Disagreed**
1	Lack of technologies /necessary equipment	18(51.43%)	6(17.14%)	4(11.43%)	1(02.86%)
2	Lack of adequate manpower (Personnel)	7(20.00%)	12(34.29%)	9(25.71%)	1(02.86%)
3	Inadequate finances	14(40.00%)	8(22.86%)	6(17.14%)	1(02.86%)
4	Nature of guideline options formulated by the political system	10(28.57%)	11(31.43%)	5(14.29%)	3(08.57%)

Total number (N) of respondents = 35.
Source: Field Survey, 2015.

Table 11.
Factors influencing technical planning and design for WEEE management systems.

were able to show that the absence of frontier technologies, essential and new equipment has hindered the operations and enthusiasm of e-waste managers. Furthermore, inadequate funding of e-waste schemes—14(40.00%) was acknowledged as a major factor influencing the ineffectiveness in the process of e-waste collection and disposal, as well as the choice adopted for final disposal measures. Collectively, 21(60.00%) policy regulators agreed that the type of guideline framed and approved by the political system sometimes militates against the effective execution of management plans for e-waste.

5. Conclusion and policy recommendations

To recapitulate, this paper discussed the transboundary movements of e-waste, the sustainability benchmarks for evaluating and adopting technologies, innovative recycling technologies, and market potential for e-waste recycling in Nigeria. With the aim of assessing the socioeconomic factors swaying e-waste generation and disposal, data collected were analyzed and discussed. The survey revealed that the structure for developing sustainable strategies frameworks and establishing resilient infrastructure for the effective management of e-waste are clearly lacking. End-users of e-waste are in the habit of stockpiling and indiscriminately disposal of e-waste. Also, it was revealed that e-waste was not segregated from household waste before final disposing. Formal recycling of e-waste is yet to be domesticated in Southeastern Nigeria. The socioeconomic reasons for the rising volume of WEEE in the study area include its cheap pricing, quality and durability, economic status of the consumer, and easy access to disused e-waste. Some of the acknowledged factors hindering the sustainable disposal of e-waste includes unavailability of innovative technologies, high cost of setting up of recycling facilities, inadequate space for stockpiling, and total obsolesce of disused EEE.

A sustainable e-waste recycling scheme would not be economically worthwhile without suitable policies in place, adoption of frontier technologies and financial measures attached. First, the management strategies for WEEE should be focused on evolving tenets of operations, and frontiers in e-waste recycling that deploys innovative and sustainable technologies. This could be achieved by adopting sustainability benchmarks for evaluating and adopting new strategies and technologies for e-waste recycling; awareness creation in the value-chain for stakeholders; as well as exploring the market potentials for e-waste recycling. These would in turn improve social and economic benefits, including decent job creations. Lastly, this can be realized through promoting appropriate policies and deliberate producer-led (and government support) initiative for recycling of e-waste.

Author details

Ojiyovwi Johnson Okorhi[1]*, Douglason Omotor[2] and Helen Olubunmi Aderemi[3]

1 National Centre for Technology Management (NACETEM), Enugu, Nigeria

2 West Africa Institute for Financial & Economic Management, Lagos, Nigeria

3 Department of Management & Accounting, Obafemi Awolowo University, Ile-Ife, Nigeria

*Address all correspondence to: johnsonokorhi@gmail.com

References

[1] Ongondo Francis O, Williams Ian D. In: Kumar S, editor. Are WEEE in Control? Rethinking Strategies for Managing Waste Electrical and Electronic Equipment. Integrated Waste Management—Volume II. Rijeka, Croatia: InTech; 2011. pp. 361-380. ISBN: 978-953-307-447-4. Retrieved from: http://www.intechopen.com/books/integrated-wastemanagement-volume-ii/are-weee-in-control-rethinking-strategies-for-managing-waste-electrical-andelectronic-equipment

[2] BCCC-Nigeria & Empa. UNEP SBC e-Waste Africa Project: Building Local Capacity to Address the Flow of e-Wastes and Electrical and Electronic Products Destined for Reuse in Selected African Countries and Augment the Sustainable Management of Resources through the Recovery of Materials in e-Wastes. In: Contribution to Components 1 and 2: Nigeria e-Waste Country Assessment; Ibadan/Nigeria and St. Gallen/Switzerland; 2011. Retrieved from: http://ewasteguide.info/files/Ogungbuyi_2012_BCCC-Empa.pdf

[3] Ketai H, Li L, Wenying D. Research on recovery logistics network of waste electronic and electrical equipment in China in industrial electronics and applications. In: ICIEA 2008—3rd IEEE Conference on Industrial Electronics and Applications; 2008. pp. 1797-1802

[4] Osibanjo O, Nnorom IC. Electronic waste (e-waste): Material flows and management practises in Nigeria. Waste Management. 2008;**28**:1472-1479

[5] Goosey M. End-of-life electronics legislation—An industry perspective. Circuit World. 2004;**30**(2):41-45

[6] National Environmental Standards and Regulations Enforcement Agency, NESREA. The National Environmental (Electrical/Electronic Sector) Regulations S.I. No. 23 of 2011, Federal Republic of Nigeria Official Gazette No. 50 Lagos—25th May, 2011. Lagos, Nigeria: The Federal Government Printer; 2011. FGP75/72011/400(OL47)

[7] Basel Convention. Where are WEEE in Africa? Findings from the Basel Convention. E-waste Africa Programme. Secretariat of the Basel Convention (SBC). 2011. pp. 1-50. Available from: http://www.basel.int/ [Accessed: January 04, 2013]

[8] The Guardian. Domestic Consumption Fuels: Africa's e-Waste Imports, Says Report. The Environment. The Guardian, Monday, February 13, 2012. 2012. pp. 48-49. Available from: www.ngrguardiannews.com [Accessed: February 13, 2012]

[9] Okorhi OJ, Amadi-Echendu JE, Aderemi HO, Otejere J. Technology paradigm for e-waste management in South-Eastern Nigeria. In: Proceedings of the 24th International Conference on Management of Technology (IAMOT 2015) Holding at the Westin, Cape Town, South Africa; 2015. Retrieved from: http://iamot2015.com/2015proceedings/documents/P099.pdf

[10] Johnson OO. Assessment of waste electrical and electronic equipment management strategies in South Eastern Nigeria [Doctoral thesis]. Nigeria: Institute of Engineering, Technology, and Innovation Management, University of Port Harcourt; 2015. pp. 1-180

[11] Eva P. Re-defining the concepts of waste and waste management: Evolving the theory of waste management [Academic Dissertation to be presented with the assent of the Faculty of Technology]. Oulu, Finland: University of Oulu; 2002. ISBN: 951-42-6821-0

[12] Pongrácz E, Phillips PS, Keiski RL. Evolving the theory of waste management—Implications to waste

minimization. In: Pongrácz E, editor. Proceedings of the Waste Minimization and Resources Use Optimization Conference; June 10, 2004; University of Oulu, Finland. Oulu: Oulu University Press; 2004. pp. 61-67

[13] Olusegun AO. Assessment of the flow and driving forces of used electrical and electronic equipment into and within Nigeria [Master thesis]. Environmental and Resource Management, BTU Cottbus; 2011. pp. 1-104. Available from: www.isp.unu.edu/publications/scycle/files/master-thesis-olusegun.pdf [Accessed: January 21, 2019]

[14] StEP. StEP Annual Report 2010. StEP, Secretariat c/o United Nations University Institute for Sustainability & Peace (UNU-ISP), Germany. 2011. Available from: www.ehs.unu.edu/file/get/8661 [Accessed: November 26, 2012]

[15] Öko-Institut and Green Advocacy Ghana. Socio-Economic Assessment and Feasibility Study on Sustainable e-Waste Management in Ghana. Report Commissioned by the inspectorate of the Ministry of Housing, Spatial Planning and the Environment of the Netherlands (VROM-Inspectorate) and the Dutch Association for the Disposal of Metal and Electrical Products (NVMP). Freiburg/Germany & Accra/Ghana: Institute for Applied Ecology and Green Advocacy Ghana; 2010

[16] Okorhi OJ, Olamade O, Aderemi HO, Obaze I, Abia-Bassey N. Strategies for sustainable management of solid wastes: A case study of industrial and commercial processes in Delta State. In: 11th Annual Nigerian Materials Congress (NIMACON 2012), Ile-Ife; November 20–24, 2012. www.msn-ng.org. Book of Abstract: Paper Code: P4-09144. pp. 140-148

[17] Federal Environmental Protection Agency (FEPA). National Policy on Environment. Abuja: Federal Environmental Protection Agency; 1999. pp. 34-35

[18] Okorhi OJ, Amadi-Echendu JE, Aderemi HO, Uhunmwangho R, Agbatah OB. Solving the waste electrical and electronic equipment problem: Socio-economic assessment on sustainable E-waste management in South Eastern Nigeria. International Journal of Environmental Technology and Management. 2017;**20**(5/6): 300-320. Available from: http://www.inderscience.com/info/ingeneral/forthcoming.php?jcode=ijetm

[19] Rolf W, Heidi O, Deepali S, Max S, Heinz B. Global perspectives on e-waste. Environmental Impact Assessment Review. 2005;**25**:436-458

[20] Basel Convention. Report of the Conference of the Parties to the Basel Convention on the Control of Transboundary Movements of Hazardous Wastes and Their Disposal on the work of its Eleventh Meeting, Geneva, 28 April–10 May 2013. 2011. Available from: www.basel.int/Portals/4/download.aspx?d=UNEP-CHW.11-24...pdf [Accessed: November 20, 2013]

[21] United Nations Development Programme (UNDP). Sustainable Development Goals (SDGs). 2017. Retrieved from: www.undp.org/content/dam/undp/library/corporate/.../SDGs_Booklet_Web_En.pdf

[22] Botkin DB, Keller EA. Environmental Science: Earth as a Living Planet. 2nd ed. New York: John Wiley and Sons, Inc; 1997. pp. 572-593

[23] Mckinney RWJ. Technology of Paper Recycling. London: Blackie Academic and Professional; 1995

[24] Peter S, Wehrle K, Christen J, SKAT. Conceptual framework for municipal solid waste management in low-income countries. Working Paper

No. 9. UNDP/UNCHS/WORLD BANK-UMP/Swiss Agency for Development and Cooperation (SDC). SKAT (Swiss Centre for Development Cooperation in Technology and Management); 1996. pp. 1-55

[25] African Business Review. How to Enter Nigeria's Booming Consumer Market—African Business Review. May 15, 2012. Available from: http://www.africanbusinessreview.co.za/money_matters/how-to-enter-nigerias-booming-consumer-market [Accessed: February 19, 2015]

[26] Nduneseokwu CK, Qu Y, Appolloni A. Factors influencing consumers' intentions to participate in a formal e-waste collection system: A case study of Onitsha, Nigeria. Sustainability. 2017;**9**(6):881. DOI: 10.3390/su9060881

[27] Basel Convention. Rules of Procedure for Meetings of the Conference of the Parties to the Basel Convention on the Control of Transboundary Movements of Hazardous Wastes and their Disposal. Switzerland: Secretariat of the Basel Convention (SBC). 2011. pp. 1-16. Retrieved from: http://www.basel.int

[28] Schluep M, Hagelueken C, Kuehr R, Magalini F, Maurer C, Meskers C, et al. Recycling - from e-Waste to Resources, Sustainable Innovation and Technology Transfer Industrial Sector Studies. Paris, France: Empa, Umicore, UNU; 2009

[29] Rochat D, Rodrigues W, et al. India: Including the existing informal sector in a clean e-waste channel. In: Proceedings of the Waste Management Conference (WasteCon2008); Durban, South Africa; 2008

[30] Krejcie RV, Morgan DW. Determining sample size for research activities. Educational and Psychological Measurement. 1970;**30**:607-610